"What I love about this book is how easily Turrell forgives us for having forgotten the promise of fusion. Generously, he takes us by the hand and brings decades of careful scientific research to life, even as the scientists whose work he traces bring fusion to the cusp of fruition."

—Gretchen Bakke, author of *The Grid: The Fraying Wires Between Americans and Our Energy Future*

"Thoroughly researched yet accessible . . . A thought-provoking read for anyone interested in learning about the cutting-edge technology that's being applied to solving the climate crisis."

—*Library Journal*

"Arthur Turrell's new book, *The Star Builders*, gives an excellent overview of the nascent international fusion industry. Turrell is himself a nuclear physicist, which enables him to analyze recent developments with an expert eye."

—CommonwealthMagazine.org

"Visionary thinkers have sketched a future of sustainable abundance based on skillful use of nuclear fusion, the process that powers the stars. Can we get there? How? When? *The Star Builders* surveys this vibrant frontier of science and technology clearly and realistically. It brings a timely, hopeful message."

—Frank Wilczek, winner of the Nobel Prize in Physics

"Incredibly readable and entertaining. The book's firsthand accounts of what is occurring inside fusion start-ups are especially enthralling. Turrell skillfully tells the fascinating story of the personalities, science, and technology that have brought this fledging industry to the point of takeoff."

—Jason Parisi, coauthor of
The Future of Fusion Energy

"Painstakingly researched. Turrell shows us a galaxy of effort being directed toward 'building a star.' Not just multinational behemoths with monster budgets but relatively small start-up companies are working with conviction, insight, and ingenuity toward a common goal: using nuclear fusion to generate near-limitless clean electricity and, with it, to save the world."

—James Mahaffey, PhD, author of
Atomic Adventures

"Whether you believe fusion will save the planet or not, everyone should know about it. If you don't, *The Star Builders* is a great place to start."

—Daniel Clery, author of *A Piece of the Sun:
The Quest for Fusion Energy*

"Arthur Turrell captures the excitement of the race to produce the first commercial fusion energy—perhaps the most important technological race of all."

—Sir Steven Cowley, director of the
Princeton Plasma Physics Laboratory

THE STAR
BUILDERS

NUCLEAR FUSION AND THE RACE
TO POWER THE PLANET

ARTHUR TURRELL

SCRIBNER

New York   London   Toronto   Sydney   New Delhi

Scribner
An Imprint of Simon & Schuster, Inc.
1230 Avenue of the Americas
New York, NY 10020

First Scribner trade paperback edition August 2022

SCRIBNER and design are registered trademarks of The Gale Group, Inc.,
used under license by Simon & Schuster, Inc., the publisher of this work.

For information about special discounts for bulk purchases,
please contact Simon & Schuster Special Sales at 1-866-506-1949
or business@simonandschuster.com.

The Simon & Schuster Speakers Bureau can bring authors to your live event.
For more information or to book an event, contact the Simon & Schuster Speakers
Bureau at 1-866-248-3049 or visit our website at www.simonspeakers.com.

1   3   5   7   9   10   8   6   4   2

Library of Congress Cataloging-in-Publication Data

Names: Turrell, Arthur, author.
Title: The star builders : nuclear fusion and the race to power the planet / Arthur Turrell.
Description: First Scribner hardcover edition. | New York : Scribner, 2021.
| Includes bibliographical references and index.
Identifiers: LCCN 2020050938 | ISBN 9781982130664 (hardcover) | ISBN
9781982130671 (paperback) | ISBN 9781982130688 (ebook)
Subjects: LCSH: Controlled fusion. | Renewable energy sources. | Nuclear energy—
Environmental aspects.
Classification: LCC QC791.735 T87 2021 | DDC 539.7/64—dc23

LC record available at https://lccn.loc.gov/2020050938

ISBN 978-1-9821-3066-4
ISBN 978-1-9821-3067-1 (pbk)
ISBN 978-1-9821-3068-8 (ebook)

For Alice

CONTENTS

CONTENTS

PROLOGUE

A CRAZY IDEA

"We need a massive amount of research into thousands of new ideas—even ones that might sound a little crazy."
—*Bill Gates, talking about ways to solve the* **energy crisis**[1]

My star-building adventure starts at a restricted-access nuclear facility fifty miles east of San Francisco's Financial District. The building I'm in is the size of three football fields—American football fields, naturally. It's called the National Ignition Facility, or NIF. Within, scientists are pushing matter to its limits, re-creating the conditions and reactions that happen inside stars.

My guide is Dr. Bruno Van Wonterghem, the operations manager for NIF. Back in 1998, he helped design the enormous facility in which we now stand. (It would be over a decade before the machine created at this site was turned on.) What he and the other NIF scientists have built here under the Californian sunshine is the world's biggest and highest energy laser.

I'm here to see what happens when this huge laser fires. It's so powerful that it can only be used in brief spurts, momentary pulses each of which the scientists call a "shot." As I wait nervously with Bruno, the director of this particular shot enters, his radiation leak detection badge swinging from his shirt, and calls us to the control room.

1

As we enter the darkened space, I see that it's laid out like NASA's ground control: computer terminals encased in plastic arranged in curved rows; operators peering at graphics, lines of computer code, and scrolling text. Perhaps Bruno knows what I'm thinking because he tells me that the laser system I'm viewing is at least as complex as the space shuttle, with 5 million lines of computer code running the show. In front of us, screens cover the wall. Bruno sees me look at the color-filled panels. "That's the list of checks that have to be completed."

An hour earlier, screens dotted around the facility began to show a colored map of the laser system. Messages sent out over the public address system instructed people to "leave laser bay 1 now, I repeat . . ." and, later, to "leave laser bay 2." In an abundance of caution, each bay was then carefully swept to ensure it was empty of human life. As each inspection was completed, the map of the facility began to show red segments indicating that this or that area was ready. The most complex shots take thirty hours of preparation.

In the control room, I glance behind me and see a couple of people looking nervously through the Plexiglas. One, a man in his mid-thirties with stubble, paces up and down. It's his experiment, and given that the typical target costs $100,000 to $150,000, you can understand why he's anxious. The National Ignition Facility, which is hosted at America's Lawrence Livermore National Laboratory, conducts a range of experiments relating to science and national security. Today's shot is classified, so I don't find out too much about what the experiment *is*—apart from that it concerns the survivability of materials that are subject to extreme radiation. What I'm *really* here to learn about are the shots aimed at creating a tiny star in the reactor chamber.

As I wait, it's almost silent apart from the crackle of a walkie-talkie as one of the twenty or so scientists communicates with colleagues across the building. Bruno says that they're charging the power supply, the world's largest capacitor bank, essentially a ginormous fast-release battery. When set off, it will loose four hundred megajoules of energy, equivalent to lighting four hundred sticks of dynamite.

The shot director begins the countdown, "Ten . . . nine . . . ," and the smaller rectangles on the big screens begin to flip from grays and reds to greens. "Eight . . . seven . . . six . . . ," intones the shot director through the public address system, and I see the biggest red bar, for the capacitor bank, begin to fill with green. "Sometimes there's a last-minute failure or a power supply fails and everything shuts down," Bruno says, and I anticipate a sudden failure, with all of the greens going to gray. But the countdown continues: ". . . five . . . four . . . three . . . two . . . one . . . SHOT!"

In the cable-choked master oscillator room behind me, a short beam of infrared light is created. It's just six meters (approximately twenty feet) long, equivalent to flipping a light switch on for twenty nanoseconds. It contains only a single nanojoule of light energy, nothing when you consider that it takes a whole joule—a billion times more—to lift an apple one meter (approximately three feet).

The fledgling beam is split into two sets of twenty-four smaller beams by precisely engineered optics. Each set travels down optic fibers to two bays of amplifiers that run almost the length of the building. There, the beams get a 10 billion times boost in energy from another laser. This is merely the start. Again, the light is split, now into a total of 192 different beams traveling in parallel. Lenses are used to expand the width of each beam to the size of an adult's chest. Any smaller, and the sheer intensity of the light would irrevocably damage the glass as the beams passed through it.

The capacitors, brimming with electrical energy and each weighing eleven tons, reach a peak power of a terawatt—greater than the entire US national grid—as they fire more than seven thousand xenon flashlamps. The lamps are similar to those used by photographers, except here, where extremes are common, they're stacked two meters (six and a half feet) high and cooled by nitrogen. The laser beams arrive and the flashlamps fire, streaking bright white light into huge slabs of neodymium-doped phosphate glass (it looks just like ordinary glass except for its pinkish tinge). As the glass absorbs the energy, the neodymium atoms within it enter an unstable, excited state. As the infrared laser beams sweep through the glass, the excited atoms relax, firing out their own infrared light. The effect is to amplify the surging laser beam many, many times over.[2]

Mirrors, each minutely deformed by little motors, reflect the incoming beams perfectly evenly so that they go back and forth through the flashlamps for yet more amplification. Start to finish, the initial beam energy is boosted 4 million billion times over. Of the four hundred megajoules in the capacitors, less than 1 percent make it into the laser beam; nevertheless these 192 infrared beams comprise, by a wide margin, the most energetic laser pulse ever created.

The beams follow mirrored paths up, around, and along the building. They pass around the control room with Bruno and me inside. They're heading for the target chamber, their 1.5-kilometer (just under five thousand feet) trip nearing its end. Each beam passes through two giant slabs of crystal. The first changes the color of the incoming laser light, converting much of the infrared to a vibrant green. The two colors pass through a second slab of crystal where they mix like paint to make a third color of light. This new beam is packed with high-energy ultraviolet photons that are beyond our visual range.

The beams enter the target chamber, a ten-meter-radius (approximately thirty feet) sphere of ten-centimeter-thick (approximately four inches) aluminium panels rounded off with thirty centimeters (approximately one foot) of concrete. Inside, it's mostly empty space kept under a strong vacuum. The beams enter, and each is focused to a round spot the diameter of a human hair.

On an *unclassified* star-creating experiment, a cylinder of gold sits in the center of the chamber. The 192 laser beams are aimed at its ends, 96 on each, to a precision of fifty-millionths of a meter. This level of precision is like trying to hit a dartboard's bull's-eye from a kilometer (approximately 0.6 miles) away. The gold cylinder is nine millimeters (a little over one-third of an inch) long and chilled to an incredible 19 degrees above absolute zero (-254 degrees Celsius)—colder than the surface of Neptune. The beams, entering through the ends, hit the cylinder's inner gold walls. The timer starts, and improbable though it sounds, what happens in the subsequent twenty nanoseconds could change the world.

In the first eight nanoseconds, the beams pack so much energy into the cylinder walls that the gold atoms are ripped apart. The ionized gold pumps the energy straight back out as yet another type of light—X-rays. Unlike laser light, which moves in lockstep, the X-rays race off in every direction.

The bath of X-rays fills the cylinder, and fourteen nanoseconds in, they reach a capsule at its center. The capsule is about the size of the pupil in your eye and so perfectly spherical that if it were as big as the Earth, the largest imperfection would be just 10 percent of the height of Mount Everest. Making a sphere so small and so perfect took hours of dexterous work with futuristic tools. The outer layer of the capsule is, incredibly, made of diamond. There's a middle layer of cold, solid hydrogen, and an inner layer of gaseous hydrogen. X-rays vaporize the outer layer, pushing hot material away from it. Just as a rocket expels hot material in one direction to move in the opposite direction, the rapid vaporization of the outer layer in one direction forces the capsule to contract. The speed is dramatic; the collapse of the capsule proceeds at a pace in excess of 350 kilometers per second (approximately 800,000 miles per hour).

The solid layer of hydrogen and its gassy center accelerate inward on themselves. They eventually reach just a thirtieth of the original capsule radius; it's as if the Earth shrank to a ball 260 miles across—much like trying to squeeze a soccer ball down to the size of a pea. The solid layer of hydrogen becomes so tightly squeezed that a teacup full of it would have a mass of over 200 kilograms (approximately 440 pounds).

In the capsule's gas center, the implosion ratchets up the temperature. The atoms in the capsule may be different, but the temperatures, pressures, and densities are similar to those found in the Sun: a tiny star has been lit. The pressure alone is 300 billion times what we experience on Earth. At the temperatures within the capsule, ripped apart hydrogen atoms crash into one another so energetically that their nuclei begin to react. But not in the chemical reactions that you might have seen in school science classes. These are nuclear reactions, nuclear *fusion* reactions.

Nuclear fusion reactions are very special. They're perhaps the most important reactions in the universe: they're what fills it with light.

Just as in stars, in each fusion reaction the nuclei of atoms are squeezed together, coalesce, and give birth to new atoms. As they do this, they also unleash vast quantities of energy. At NIF, hydrogen nuclei combine to make helium nuclei. The fusion-released energy manifests in the frenzied speed of the helium, each nucleus rushing out at 13 million meters (a little more than 42 million feet) per second. As the outgoing nuclei crash into surrounding hydrogen nuclei, they heat those nuclei up, increasing the chances that more hydrogen will react, making more fast helium nuclei, and so on.

Nearly twenty nanoseconds after the lasers have entered the cylinder, in excess of 10 million billion reactions have happened. Each one turns matter, the stuff we're all made of, directly into energy.[3]

Eventually, the ball of hot fusion fuel is unable to remain whole. It's not natural to have an object 0.1 millimeters (approximately four thousandths of an inch) across, as hot and as high pressure as the Sun's core, fusing in the center of a ten-meter (a little more than thirty-two feet) vacuum chamber. The fuel quivers and wobbles; it's only held in place briefly by its own inertia. Sound waves, positively sluggish compared to the particles, travel through it, breaking it up. The reactions are over. The star is dead. At least, until the next shot from the laser, and the next target.

"Not very spectacular," Bruno says after the shot is over. I don't believe him, not even for a nanosecond.

Star Power

The idea of building a small slice of star matter on Earth has, since the 1940s, captivated scientists, governments, billionaires, entrepreneurs, celebrities, a pornography magnate, and even a few dictators. The scientists of the National Ignition Facility aren't alone; for years, groups all around the world have been devising elaborate star machines that stretch human ingenuity to the breaking point. Now those star machines are being built and operated. The machines range from contraptions cobbled together on a shoestring to a seven-storey mechanical doughnut rising out of the French countryside. The star-trapping tricks they use are straight out of science fiction: force fields, lasers, and pneumatic pistons. The teams that build them are comprised of physicists, engineers, mathematicians, and computer scientists who've dedicated their entire careers to containing and controlling the nuclear forge that is a fusion reaction. They are the star builders.

At best, creating a mini-star on our home planet sounds highly inadvisable. At worst, it sounds like the diabolical plan of a villain from a James Bond or *Star Wars* film. In this book, I'm going to show why the star builders' ideas, crazy as they might seem, could actually save the planet, and who's ahead in the race to control and exploit star power.

The real goal of the star builders isn't to re-create a star exactly, but to re-create and control the *power source* of stars—nuclear fusion— here on Earth. Nuclear fusion is different from nuclear fission, the reaction that occurs in today's nuclear power plants. In nuclear fission reactions, large nuclei—the cores of atoms—are ripped apart into smaller nuclei. In contrast, nuclear fusion creates larger nuclei out of two smaller nuclei. Both types of reaction get their name because they happen to the core of an atom, the nucleus. But their risks, the amount of energy they release, the nuclei that will undergo them, and the technologies needed to make them happen are very different.

Controlled nuclear fusion is just the last, unconquered part of a quartet of nuclear technologies. The other three are controlled fission as used in today's nuclear reactors, uncontrolled fission as used in atomic bombs, and uncontrolled fusion as unleashed in thermonuclear, or hydrogen, bombs. All of the other three were demonstrated in the 1940s and 1950s. We still don't know how to do the fourth, but the star builders are trying hard: and they say that they're getting close.

What does it mean to control, or achieve, nuclear fusion? For star builders, it's not just about making a few atoms fuse together to form bigger atoms. Actually, as nuclear physics goes, that's easy enough. Plenty of people have done it already, whether in laboratories with thousands of staff or in the school classroom, like the boy who trawled eBay for parts to build a fusion machine. But the amount of energy generated by these rigs is piffling compared to the energy used to make them work.[4]

Star builders are trying to show that fusion can produce *more* energy than it uses, that fusion is a viable power source. Producing more energy from reactions than it takes to get those reactions going in the first place is the first step. "Breakeven," "net energy gain," or the self-perpetuating "ignition"—star builders use lots of terminology, but it all means "more energy out than in." And teams of star builders around the world are racing to be the first to do it.

Net energy gain is exactly what you get when you light a log with a match; it doesn't take much energy to light a match, but the roaring fire it ignites releases that energy many times over as heat and light. Star builders often talk about this in percentage terms, with 100 percent being breakeven and greater than 100 percent being net energy gain.

But getting to 100 percent, as much energy out as in, is an enormous scientific and technical challenge. No one has yet done it. And that feat is the key to unlocking fusion as a power source. Once star builders have shown that fusion can deliver just as much energy out as was put in, it's just a matter of optimization to get even more energy out. Only then will fusion be able to change the world. It's the difference between the Wright brothers' plane not flying at all, and its flying 250 meters (approximately 820 feet). Making the leap from flying 250 meters to flying for miles is easier than being rooted to the ground and figuring out how to fly. Psychologically, getting to the first level of accomplishment is everything.

Eventually, star builders want to achieve a gain of 3,000 percent (thirty times energy out for energy in), or even 10,000 percent (one hundred times). To the star builders, really achieving fusion means creating a genuine power source based on the nuclear reaction that keeps the Sun shining.

This is phenomenally difficult. If I say that controlling fusion to produce energy is the biggest technological challenge that we've ever taken on as a species, it will sound like hyperbole. But it's true. Fusion at NIF needs, first, temperatures in the hundreds of millions of degrees, and second, matter as densely squished as the material in the Sun's core. The complexity of the machines is beyond anything we've ever designed. There are *tens* of millions of indi-

9

vidual parts to a star machine. NASA's space shuttle had just 2.5 million. I'll keep returning to the space analogy because, honestly, there are few other close comparisons when it comes to the scale of the challenge—and there are few environments anywhere near as extreme as those in star machines.[5]

There are two practical ways to create the magic conditions that make fusion happen. One is called *magnetic confinement fusion* and the other is *inertial confinement fusion*. There's gravity too, of course, but for that you need scales bigger than can be created on Earth: you need, quite literally, a star. The magnetic approach is to bind the hot matter in a reactor with an invisible web of magnetic fields. The inertial approach sets matter crashing into itself, thereby both heating and compressing it, and aims to get all the fusion done before the assembled star matter falls apart again. NIF uses lasers to do this.

You may wonder why anyone would bother trying to re-create the fusion taking place in stars. Is it sheer arrogance? Such attempts to dominate nature can sound like human folly. It's true that star builders are partially attracted by the sheer challenge. The theoretical physicists and computer scientists want to know if their models and simulations can come close to reality. The experimentalists want to understand and measure untold extremes. The engineers want to build machines that can withstand such extremes. But there's another reason too, one that's much more important for the rest of us. Building a star and perfecting power from nuclear fusion could provide humanity with millions, perhaps billions, of years of clean energy. When the BBC asked Professor Stephen Hawking what world-changing idea he would like to see humanity implement, he said, "the development of fusion power to give an unlimited supply of clean energy." No wonder star power is sometimes described as the "Holy Grail of energy production."[6]

Like the Holy Grail, fusion has been elusive. This has spawned a delightful range of jokes. Most star builders have their favorite, but perhaps the most pointed is "Fusion is the energy of the future . . . and always will be," with common variations along the lines of "Fusion is thirty years away . . . and always will be." The British prime minister had his own go in 2019, saying of a European Commission–run fusion lab: "They are on the verge of creating commercially viable miniature fusion reactors for sale around the world. Now I know they have been on the verge for some time. It is a pretty spacious kind of verge."[7]

Certainly, it wasn't meant to be this way. "It is the firm belief of many of the physicists actively engaged in controlled fusion research in this country that all of the scientific and technological problems of controlled fusion will be mastered—perhaps in the next few years," said early star builder Richard Post, a scientist at the predecessor of Lawrence Livermore National Laboratory. That was in 1956.[8]

For a long time, the commitment of a scientist to fusion was measured in decades rather than years. The tribe of star builders around the world transcends language, political leaning, and cultural differences. A good example is how, at the height of the Cold War, a British team went over to Russia to find out about and verify the then astonishing progress of a new, Soviet-designed magnetic confinement fusion device. Or how China, Russia, the US, the EU, India, South Korea, and Japan are working together today on the direct descendent of that Soviet machine. Even with an unusual spirit of collaboration, government backing, and very talented people, great patience has been required. As lifelong star builder Marshall Rosenbluth put it in 1985, "May our grandchildren live to see fusion power."[9]

Perhaps one of the issues making nuclear fusion difficult is its name. Say that a technology involves "nuclear" anything and you may find yourself getting a frosty reception. You can see why nuclear has a bad reputation in some quarters. Most commonly, it's associated with nuclear *fission* power, which divides opinion and produces radioactive waste that we'll have to store for thousands of years. At

worst, it's associated with the death and destruction wrought by nuclear weapons. Star builders have a lot to say about why nuclear is
not automatically bad. Nuclear is a tool like any other technology.
We've seen its worst already, they say; now it's time to see its best.

Their arguments that nuclear fusion could help save the planet
have stirred interest among the public. Sir Steve Cowley, a professor who now runs a star machine at Princeton University, has given
a talk on fusion that has received more than half a million views;
Taylor Wilson, who built his first fusion reactor at fourteen, has a
TED Talk that's racked up millions of views.[10] Some fusion start-
ups have tapped into the interest directly through crowdfunding
websites.[11] Even Hollywood has caught the fusion buzz: both Bat-
man and Spider-Man have grappled with evil star machines.

Whether all the talk about fusion has propelled fusion research
or, rather, scientific breakthroughs have increased the chatter,
there's no question that the race to achieve fusion is heating up.
There's a wave of new or overhauled star machines coming online.
There are more competing teams than ever. There's been a sudden
rise in the number of private fusion start-ups, creating the biggest
shake-up in star building since the 1960s. Despite the risks of dealing with matter at extremes, investors are betting that private firms
can succeed where governments have failed. By 2019, the new fusion start-ups had received funding in excess of $1 billion.[12]

"An energy miracle is coming, and it's going to change the world,"
says Bill Gates, who has sunk some of his cash into one fusion start-
up. PayPal founder, billionaire, and member of former President
Trump's inner circle Peter Thiel has done the same, investing $1.5
million in 2014.[13] Other Silicon Valley entrepreneurs involved in
fusion include Jeff Bezos, executive chairman of Amazon, and,
before his death in 2018, Microsoft cofounder Paul Allen.[14] Goldman Sachs has plowed money into an effort, Lockheed Martin
has its own initiative, and even fossil fuel firms like Chevron are

hedging their bets by investing in fusion. In a sign of the times, one fusion scheme is backed by Brad Pitt; another by a reality TV star, Richard Dinan of the UK's *Made in Chelsea*. Dinan has posted pictures on social media of what looks to be a photo shoot inside a reactor vessel ("Just chilling in my fusion reactor").[15]

Such is the activity that a fusion industry lobby was recently formed and the US House of Representatives passed a bill unlocking hundreds of millions of dollars of funding for public-private fusion partnerships.[16] It's not just eccentric Silicon Valley entrepreneurs funding fusion; Legal & General Capital, the early-stage investment arm of Europe's second largest institutional investor, is making "modest" investments in nuclear fusion. Senior investment analyst at Legal & General Nicola Daly said, "Nuclear fusion has the potential to be a game-changer in a world that needs some game-changers."[17] But it goes bigger still: the governments of Canada, Malaysia, and Russia are hedging their bets by putting cash into challenger fusion firms alongside their existing science programs.

The British PM may have joked that fusion was taking longer than expected, but his speech also announced an additional £200 million (approximately $270 million) of funding. The Obama administration's science advisor argued that fusion needs much more investment. In 2016, the German chancellor Angela Merkel, who also holds a PhD in quantum chemistry, launched a new experimental fusion reactor with the words "Every step we are taking on the long road towards a fusion power plant is a success."[18]

As well as the international experiments that are in progress, governments around the world are beginning to pursue home-grown fusion reactors. International collaborations can be slow because they become mired in discussion about where facilities will be built (and so who receives the benefits of investment) and who will build their components. Running a parallel in-house scheme is expensive but could allow a nation to leapfrog the competition.

"Currently, megajoule scale lasers are under construction in both France and Russia. The Chinese have completed and are operating the second most energetic laser in the world and are publishing papers with designs for lasers fifty percent to three times the size of NIF," testified Dr. Mark Herrmann, director of NIF, to the US House of Representatives in 2018.[19]

China will begin construction of a net energy gain–capable magnetic fusion device in Hefei in the 2020s, using the knowledge they have honed on a smaller but similar machine called EAST. They also have an inertial confinement site called the Shenguang-III laser facility. A Russian news agency released a video in 2019 confirming their government was building a facility to rival America's National Ignition Facility, although details have been scarce. In 2020, there were eighty-eight fusion reactors in operation around the world, with a further nine under construction. Public and private, big and small, star machines are taking off.[20]

Everyone is talking about the new wave of innovation.[21] One estimate by the International Atomic Energy Agency suggests that the annually published number of peer-reviewed research papers on fusion for energy has more than tripled since the mid-1990s. Net energy gain used to be decades away, and always would be. Now both start-ups and national laboratories are saying that net energy gain is a question not of if, but when. Not decades, but years. Not how, but who—who will get there first? In this book, we're going on a journey to find out why fusion is so important to the Universe, how it could be transformative for planet Earth, and who is closest to taming its tremendous energy. Whoever does achieve net energy gain first is going to fundamentally change how fusion is seen. Just as occurred with flight, once "fusion for energy" is clearly demonstrated, an explosion of innovation will be unleashed. And from there the path may open to powering the planet.

CHAPTER 1

THE STAR BUILDERS

"If, indeed, the sub-atomic energy in the stars is being freely used to maintain their great furnaces, it seems to bring a little nearer to fulfilment our dream of controlling this latent power for the well-being of the human race, or for its suicide."

—Arthur Eddington, "The Internal Constitution of the Stars," 1920[1]

Who are the fusion pioneers aiming, like Prometheus, to steal the secret of fire from the heavens? The individuals who are bold enough—some might say "crazy enough"—to try to bring star power to Earth? Throughout this book we'll be meeting them and learning why they've dedicated their lives to the fusion dream.

The first star builder I meet as I try to find out who is ahead in the nuclear race is Dr. Mark Herrmann, the gentle-mannered director of the National Ignition Facility (NIF) based at Lawrence Livermore National Laboratory.

Like everyone I meet here at NIF, Mark opens our conversation by stressing that managing the United States' stockpile of nuclear weapons is the primary mission of both NIF and Lawrence Livermore National Laboratory. The scientists here are tasked with maintaining America's nuclear deterrent and understanding how aging nuclear weapons deteriorate over time. This is why the entire site is

15

protected by armed guards and lined with serious-looking double fences. As I walked along Livermore's winding paths to get to my meeting with Mark in the NIF visitors' center, I passed numerous other buildings that were strictly no entry for those without security clearance. Inside, the weapons secrets of the most powerful nuclear state on Earth are held. The combination of high security and the brightly colored visitors' center might seem incongruous, but everyone I talk to is friendly and seems to have found peace with their responsibilities, Mark especially.

Livermore does a vast range of science in addition to weapons research and nuclear fusion; super-computing, climate change, and the creation and discovery of new elements (including livermorium, which is named after the lab). Make no mistake, this is big science—NIF alone has 650 staff who are managed, ultimately, by Mark. I begin by asking him how close the lab is to demonstrating net energy gain.

"By the end of the 2020s we'll have achieved ignition or have an ignition facility under construction," he says, his eyebrows jumping above the rim of his thick glasses to make the point. "Ignition" means a high net energy gain from nuclear fusion in which the reactions really take off and become self-sustaining, like a roaring fire. Mark has been working on unlocking energy from atoms for more than two decades, and leading NIF since 2014. Although he has graying hair and a salt-and-pepper goatee, he's full of energy and enthusiasm for NIF's mission.

Mark was previously employed at the Sandia National Laboratory, where he was the director of the Z Pulse Power Facility, another machine that combines classified and open science. When I ask why he's at NIF, he tells me that he got into fusion research because of the interesting science and the long-term benefit for humanity. His first step in the field was completing his PhD in 1998 at Princeton and writing an award-winning thesis on the rival magnetic confinement approach. Shortly after, he joined Livermore to work on inertial confinement fusion.

Despite Livermore's focus on stockpile stewardship, one of the laboratory's long-term goals is inertial fusion energy, and always has been since its founding in 1952. Mark is clear that NIF is the world's best hope for understanding fusion, and he tells me that it's the only facility that has the prospect of achieving net energy gain in the next decade. That's controversial given that other fusion laboratories and start-ups are claiming that they're ahead. Earlier in the day, Dr. Bruno Van Wonterghem, NIF's operations manager, told me that the extent to which Livermore is explicitly pursuing fusion has gone through "highs and lows," perhaps hinting that the political weather might be why everyone I spoke to began our conversation by telling me the primary objective was maintaining the United States' nuclear arsenal.

I ask Mark about the tension between managing nuclear weapons and pursuing fusion energy. "Holistically it's all stockpile stewardship." What he means is that the physics of fusion reactions is similar whether those reactions are occurring in a thermonuclear weapon, in a fusion reactor, or in space.

One person in particular is representative of the strides that NIF has made since 2013, and is also most emblematic of the paradox of mass destruction and planet-saving energy provision that characterizes Livermore's portfolio: Dr. Omar Hurricane, NIF's chief scientist. You'd be forgiven for thinking that he was the star of an action film with a name like that; as it is, he's something of a star in the inertial confinement fusion community. His thesis advisor was the UK's previous star builder–in-chief Professor Sir Steve Cowley, but after Omar finished his PhD at UCLA in 1994, he left magnetic confinement fusion in favor of an inertial confinement fusion job at Lawrence Livermore.

"I got hired into the weapons program instead," he tells me, as we sit down to talk. He was irked by his rejection but made the most of the unexpected career he found himself in. After nuclear testing ended in 1992, a different type of stockpile stewardship was required. "How can we be confident about certifying the nuclear stockpile," Omar says, "when we're not doing experiments anymore? That led to the stockpile stewardship program, my generation." He was involved in extending the lifetime of the W87, a thermonuclear bomb that is used in intercontinental ballistic missiles.

Omar isn't afraid of celebrating his successes: "I'm pretty good at making mathematical models of things, even things that aren't my area," he says, and explains that, following the worse-than-expected performance of fusion experiments at NIF, "the director of the lab saw it wasn't going well and asked me and a few others from the weapons program, 'Would you be willing to jump in and help?' And so I jumped in with other colleagues. The experiments were quite successful [and in] late 2013, early 2014, we started getting some exciting results. All of a sudden, I got asked whether I wanted to be chief scientist."

Under Omar's leadership, laser fusion experiments performed at NIF in 2018 released sixty times the energy of experiments on the same machine in 2011. But NIF isn't the only front-runner with decades of experience in nuclear fusion.

Five thousand miles away, a UK government laboratory called the Culham Centre for Fusion Energy has the latest iteration in a line of fusion machines that goes back decades. It's now the world's leading operational magnetic fusion facility. Unlike Livermore or Sandia in the USA, Culham doesn't do classified weapons science. The site isn't protected by armed guards, although on my way in I did see some pretty mean-looking ducks. Star power is the sole mission.

Professor Ian Chapman leads both the laboratory and the UK Atomic Energy Authority, the arm's-length civil service organization tasked with star building. Ian Chapman is exactly what you'd expect if you crossed a scientist with a civil servant. He wears a suit and tie (unusual for scientists), but it's out of respect for the seniority of his position rather than pretense. He has close-shaved hair and a broad grin. He's thoughtful, talkative, and polite, but he's also not one to mince his words. That's useful if you're trying to steer a fifteen-hundred-person laboratory. Most of his staff are scientists, each with their own interests, and I imagine the internal management of Culham involves a degree of cat-herding. He sees his role in leading the world's largest (for now) magnetic fusion experiment as a duty, though he clearly misses being in the details.

"I'm chief executive and my role here is fundraising, stakeholder management, dealing with the government, Brexit"—he chuckles, acknowledging the scale of that particular challenge for Culham, whose biggest fusion experiment is funded directly by the European Commission—"all that not very fun stuff."

We're talking in Chapman's office, which, despite his responsibilities, looks like the inside of an office trailer on a building site. The only hints that it might not be are the equations on the whiteboard. Ian is another award-winning scientist, having bagged the latest of many trophies in 2017 for research on the stability of magnetic fusion experiments. I ask him about the prize and he's characteristically self-deprecating.

"I used to be a scientist—yeah, I just received an award for science I used to do. I spent thirteen years doing proper science, but I've written off doing any real work while Brexit is happening, as that's going to occupy me for years."

It's worth noting that the award was for outstanding *early career* research. Ian has risen remarkably fast. He went from finishing his PhD in 2008, to making groundbreaking contributions to science, to running the world's most successful fusion experiment in less than a decade.[2] Given his inexperience, his appointment was described by some as a risk.

"It's a risk in that I didn't have decades of experience running big organizations with thousands of people. Conversely, had you appointed someone who knew how to organize but didn't have a passion and a knowledge about fusion, you'd be taking a risk at the other end. It's clear that I have a passion about fusion. I'm also the right age profile to make it happen, shall we say," the thirty-eight-year-old adds, smiling.

Culham's biggest machine currently holds the world record for fusion energy. Chapman has a plan to push it even further than before. "I'm hoping we can smash our record," he has said.[3]

Established star builders like Mark Herrmann and Ian Chapman face stiff competition from elsewhere: the Cambrian explosion of private fusion firms. Like the industrialists of the early twentieth century, these challengers are less concerned with the science than in making the machines work. They reject the "bigger is better" paradigm that is conventional wisdom in fusion physics. Instead, they're developing smaller and, they argue, more practical machines. Increasingly, investors are committing their cash to this scaled-down, simplified approach—though of course no fusion device is without vast complexity.

One such efficiency-emphasizing operation is Tokamak Energy. Although Tokamak Energy's scientists and engineers are following the lead of Culham in using magnets to trap the stuff of stars, they believe their machine is a smarter way of doing it. Not only are they aiming to demonstrate that they can reach the conditions for fusion soon, they want to deliver power to the grid by 2030. This ambitious plan will involve mastering new intricacies in not just physics, but engineering and economics too. In 2020, Tokamak Energy received $13 million from the British government and another half a million dollars from the US Department of Energy to bring this plan to fruition.[4]

Fusion start-ups such as Tokamak Energy are ending the domi-

nance of physics, and physicists, in the field. Tokamak Energy's chief
executive, Jonathan Carling, is the quintessential engineer determined to turn fusion from a science project into a bona fide power
source. On the day of my visit I meet him over tea and biscuits in a
room cut out of Tokamak Energy's industrial warehouse headquarters. Jonathan is unlike the incumbents in the star-building business in that he's never worked in fusion before. But he *has* taken big,
complex engineering designs into commercial production.

"I came to be here because my background is in engineering
and operations, and the business has reached a stage where it's very
focused on how we make this a commercial reality, not just how we
demonstrate an energy gain of one-point-zero-something, but how
we actually develop a commercial device."

His record speaks for itself. His career began when he appren-
ticed at Jaguar to work on car engines, but he says that his passion
for making technology work began much earlier.

"I became an engineer when I was about six," he tells me. "We
used to get drawing to do and I drew a car and my teacher would
say 'What's that sticking out of the [hood]?' and I said, 'That's a
super-charger.' At six years old, I was fascinated by machines, and
I was always pulling things to bits and wanted to learn about them.
So I did a mechanical engineering degree."

After Jaguar, Jonathan went to another high-end car firm, Aston
Martin, and became the chief operating officer. Not content with
the complexity of cars, not to mention people, he then switched
to aerospace at Rolls-Royce. If you've ever been on a plane, there's
a really good chance that you've been jetted around by an engine
that Carling had a hand in. At my insistence, he reels off a list: the
Airbus 380, 350, and 330, the Boeing 747, 777, and 767.

"Jet engines run hot all the time, and the intake temperature
can be of the order of two thousand Kelvin, which is three hundred
degrees or so above the melting temperature of the turbine that is

extracting the power," he tells me, going on to explain how such a feat is possible with clever engineering. If it sounds impressive, it's nothing compared to what a working fusion reactor will need. So why did he swap two thousand degrees Kelvin for 150 million?

"The world doesn't need another luxury sports car as much as it needs fusion energy," he says.

There are start-ups pursuing the inertial confinement approach too, which as a reminder uses a trigger, often laser beams, to crush matter into a hot, dense blob that provides good conditions for fusion reactions to occur. One firm is on the other side of Oxford, just seventeen miles up the road from Tokamak Energy and in their own rather more swanky warehouse. They are First Light Fusion; their name refers to the light emitted by matter as it gets hot enough for fusion.

Dr. Nick Hawker, the CEO and CTO of First Light Fusion, is another engineer steering fusion toward reality. He's young, having founded First Light directly after completing his PhD at Oxford. He has a very different style from the older bureaucrats of fusion, like Mark Herrmann and Ian Chapman. Nick wears sneakers, chinos, and solid-color T-shirts with a blazer over the top. He's sharp and a bit intense, with a restrained entrepreneurial energy. Someone who decides to take on not one, but two of the key roles in an organization might be conceited. Yet, unusually, Hawker's company has managed to raise enough money for a four-year plan. And his generally older and more experienced staff look up to him with a hint of reverence. When I finally speak to him at the end of a long day at First Light's headquarters, I'm full of anticipation. Hawker answers my questions in clipped sentences and rarely cracks a smile throughout our conversation; he's all business.

He tells me that, as CEO, his job is forging links with potential industrial partners and academia. But he seems to be involved with every aspect of the company and likes getting his hands dirty. His Twitter feed is full of results from First Light's star machine, replete with video clips of experimental equipment exploding. One post shows a photograph of a thick metal plate with a hole punched right through, another a video of a 7 mega-ampere (25 million times more current than an old-fashioned filament bulb) short circuit.

I ask him about his dual role, and he says that it's all about managing the personalities on the team so that they're pointing in the same direction. "I'm on the pitch too," he quips, referring to getting directly involved in the science via his CTO role.

And well he might get involved in the science. Nick has taken his work in Oxford's engineering department, simulating extreme conditions in fluids, and made it the core of a new approach to fusion. That's risky, but it also presents new possibilities for net energy gain that, he argues, might have been overlooked by the big laboratories, who tend to play it safe.

Certainly, decades of mainstream magnetic and inertial confinement fusion have continually surprised scientists with problems that couldn't have been anticipated. But technology has also moved on, and fusion scientists can now combine simulation and theory in ways that would have been unthinkable ten years ago. "The real goal is to validate the simulations," Nick stresses, echoing what Jonathan Carling also told me. To keep up the funding to get their fusion schemes over the line, the start-ups need to show that they're credible, and that means showing that their models are capable of describing reality.

The big motivations for Nick seem to be those that have driven countless entrepreneurs before him: success and money. He believes that these will come faster in a private fusion venture.

"I'm very glad that we did go private because look how far we got," he tells me. Not only does Nick live this paradigm, he has championed entrepreneurship in the press and mentored others starting businesses. He really believes that when it comes to technological progress, his way is the best way.

Nick Hawker is counting on getting to a net-energy-gain experiment by 2024. He tells me First Light Fusion is about to reach the temperatures where fusion reactions become detectable, a first step on the path to this ambition.[5]

First Light Fusion fits into the inertial confinement fusion bucket. There are fewer start-ups using this approach, the most prominent examples being New Jersey's LPP Fusion and Canadian-based General Fusion. The other start-ups using the magnetic field approach include Lockheed Martin, TAE Technologies, Commonwealth Fusion Systems, and Renaissance Fusion. They're all looking to challenge the big players of NIF and Culham with their greater agility and more focused objectives.

It doesn't matter which star builder you talk to. All passionately believe that their scheme will be the first to deliver energy to the grid. But they can't *all* be right. Some are over-promising. And the path to fusion energy is littered with failed promises, so, in a way, it's surprising not to hear more modesty.

Despite having his own ambition to surpass records for fusion energy, Ian Chapman believes that the new competition is essential and inevitable: "I'm very supportive of all of their endeavors, and indeed we work with a lot of private companies." But he does acknowledge that the start-ups can create problems and may not be as far down the road as they think they are.

The government labs may still have a few tricks up their sleeve too. They're not incapable of innovating, and they have lots of people with the right skills to do so. Both magnetic confinement fusion, at Culham, and inertial confinement fusion, at Los Alamos National Laboratory, have smaller, highly experimental fusion schemes.

Perhaps the most impressive government "start-up" machine is the Wendelstein 7-X, recently opened by Angela Merkel in Greifswald, Germany. It's run as part of the Max Planck Institute for Plasma Physics, which has eleven hundred employees and also operates a tokamak. W7X, as those in the know call it, has been making rapid progress by revisiting an idea right from the start of the fusion era, the stellarator (an Escher-like tangle of tubes that traps fusion fuel with twisting magnetic fields), with modern technology. Serving as the institute's scientific director is Professor Sibylle Günter, an experienced academic star builder whose work on topics related to fusion began in the 1990s. It was the connection of fusion with her corner of northeast Germany that drew her in. When she discovered that W7X would be built near her hometown of Rostock, she decided to learn more.

Sibylle steadily ascended the ranks, becoming head of theory (a position for which seriously strong mathematical ability is required), then a director, and finally, in 2011, the scientific director. "I saw how important good management is and how much it takes to secure a sufficient budget," she tells me digitally as we cope with a coronavirus-induced lockdown. "By being the director I have many opportunities to influence our big projects and I can change those things I only complained about earlier."

Although she describes herself as a very impatient person, Sibylle is understated; the consummate professional scientist giving both sides of the argument and being honest about any limitations. When I ask how she feels about achieving fusion, she says, "The pressure is quite strong" but adds that, despite it, she still strives to ensure careful work and good scientific procedures. She thinks that, in the long run, the stellarator design of W7X could make a more viable energy-producing fusion reactor than tokamaks like Ian Chapman's machine at Culham.

Although everyone disagrees about *how* and *who*, star builders do agree that the fusion future we've all been promised for so long is (almost) here. Net energy gain especially. "The key message isn't about us," Jonathan Carling told me. "The key message is that fusion is coming much faster than most people think."

"It's not science fiction; it's going to be solved in the next decade," Nick Hawker said. "'Solved' means it's working. It's going to take longer for a power plant, but the joke that fusion is 'thirty years away'—no, it's here. The thirty years are done and it's going to be solved in the next decade."

Ian Chapman, who leads the UK Atomic Energy Authority, said: "Fusion *will* work. It *will* happen."

There's just one more star builder I need to introduce you to: me. Well, *former* star builder: I worked on nuclear fusion research at Imperial College until 2015, when I left to become a researcher in economics in the public sector. Since then, I've been looking at star builders from the outside. In that time it has struck me that, more than ever, the rest of the world deserves to know what this motley bunch of scientists, engineers, and entrepreneurs is up to. This book is intended to help accomplish that.

It's clear from talking to the star builders that they aren't just creating fusion devices to show that they can master one of the universe's most fundamental reactions, as important a breakthrough as that might be scientifically. They're doing it because taming nuclear fusion *might* just save the planet.

CHAPTER 2

BUILD A STAR, SAVE THE PLANET

"What problem do you hope scientists will have solved by the end of the century?"

"Nuclear fusion. It would provide an inexhaustible supply of energy without pollution or global warming."

—Stephen Hawking, 2010[1]

The star builders we've met—Dr. Mark Herrmann at NIF, Professor Ian Chapman at the UK Atomic Energy Authority, Jonathan Carling at Tokamak Energy, Dr. Nick Hawker at First Light Fusion, and Professor Sibylle Günter at the Max Planck Institute for Plasma Physics—have many motivations, but there are a few that come up again and again. One is the sheer joy of pushing the boundaries of what human ingenuity can achieve. For the start-ups, success and money beckon. For scientists in government laboratories, it's understanding the most powerful forces in the universe.

But the big motivation that unites all star builders, the one that gets them out of bed in the morning, is saving the planet.

Our home is a blue marble that sits in a thin protective layer of atmosphere. It's currently hurtling through the Milky Way, following the star that has given it life. Earth has been witness to every joy that humans have known: every sight, every experience, every moment. (That's if you don't count the lucky few who have ventured into space.)

As much as everyone agrees that we should preserve our planet, keeping it pleasantly habitable for future generations of plants and animals, the star builders are acutely aware that, as a species, we're charting a course to do it irreparable harm. "We're doing an experiment with the only ecosystem we have," Mark Herrmann told me. "Bad things very much could happen." And so they have.

Simply put, Mark and the other star builders want to stop the human-caused climate change that is tipping the planet into a new and dangerous phase. Climate change threatens our way of life, especially our most vulnerable people. That's the A-side; the B-side features the related challenges of pollution and habitat destruction.

There's clamor for change—though, so far, not a lot of concrete progress. Teenage climate campaigner Greta Thunberg is touring the world telling politicians and officials to do more. In the UK, the Extinction Rebellion movement has protested by blocking streets and gluing themselves to government buildings. They want net zero carbon emissions by 2025. In the US, Democrats have campaigned for a Green New Deal that includes a commitment to net zero by 2030. By 2019, fifteen nations had committed to reach net zero by 2050. But it's one thing to promise net zero carbon emissions, it's quite another to achieve them.[2]

The star builders have an unusual plan to avert Earth's unfolding climate catastrophe: they want to save the planet by building a star. Star power, they say, could help us avoid the worst of the planet's climate crisis by providing an alternative source of energy. And it's our relentless addiction to energy that has caused this unprecedented crisis.

No other animal uses energy like humans do. Way back in our evolutionary history, we used energy to cook food to improve its nutritional value.[3] And since *Homo sapiens* first used fuel for fires, our hunger for energy has inched higher and higher.

We began to use energy as an input into technology, from baking ancient bricks in kilns to propelling steamers across the oceans. Over the last two hundred years, the increases in living standards that energy has enabled are astounding. Mastery of energy has seen us improve the quality of hospitals, communications, transport, and both the quality and quantity of our leisure time—to pick but a few examples. The average home is packed with labor-saving devices that would have been unthinkable three hundred years ago.[4]

Each revolution in technology has meant successively higher energy consumption. In the UK, where the Industrial Revolution began, annual energy consumption increased eighty times between 1700 and 2019. Worldwide, it's up thirty times since 1810.[5]

We'll almost certainly need more energy in the future. Just as they have in the past, life-improving technologies to come will likely use more energy. And future gains in productivity, which ultimately underlie prosperity, are likely to come from energy-hungry technologies like robotics.

More energy is needed right now just to even out existing inequalities in energy consumption. A large fraction of the world's population doesn't enjoy anywhere near the extensive energy consumption, and the benefits it brings, that people in developed nations do. The hope is that poorer countries catch up as quickly as possible. When they do, the scale of the increase we'll need is huge: for example, energy consumption per person in Bangladesh is around eleven times smaller than in the UK, although the population is roughly twice the size.

The population of the planet is increasing too, especially in developing countries. Population growth gets bad press, but actually it means more brains creating more wonderful things across art and science.[6] Nigeria, Africa's most populous nation, experienced an astonishing growth rate of 45 percent between 2005 and 2019, compared to the US's 11 percent for the same period.[7] The overall rate of population growth is predicted to slow only gradually in the coming decades before stabilizing at around 10 billion people.[8] We will need energy for those new arrivals.

It's hard to avoid the conclusion that we have a big energy problem. "Frankly, any projection you look at says you need more energy," Ian Chapman told me. "We're going to need half as much energy again as we use now—50 percent more, it's a huge amount more. In this country, if we had demand go up 15 percent we couldn't cope."

Energy is measured in joules. Remember, a single joule is roughly the energy it takes to lift an apple one meter. Over the course of a year, the average US citizen uses three hundred gigajoules, with a gigajoule being a thousand million joules. There are 330 million people living in the US, so, to account for the energy use of the entire country, or an entire planet, an even more extreme unit of measurement is called for: the exajoule. A single exajoule is a thousand million gigajoules.

Today, the world uses 620 exajoules of energy per year. Of those, the US uses around 95. The US Energy Information Administration estimates that the planet will need half as much energy again by 2050. Other estimates put it a touch lower. Even for everyone now on the planet to have the same energy consumption per person as in the EU or the US, we'd need to find another 370 or 1,650 exajoules respectively. That's a lot of extra energy that we need to find—and it has to come from somewhere.[9]

The Energy Crisis

If you think that the energy crisis that worries star builders so much is being solved because campaign groups are pressuring and electorates are increasingly concerned, you're horribly wrong. While there's a lot of talk, and ambitions are high, the data tell a different story. Both the world's total and per person energy consumption are increasing, partly for the very good reason that extra energy directly makes people's lives better. Over the decade to 2018, global primary energy consumption grew by 1.4 percent a year.[10]

Most of the world's energy (80 percent) is still generated by fossil fuels: oil, gas, and coal. That's not new—throughout modern history, we've relied almost exclusively on a single type of chemical reaction for energy. The recipe: take hydrocarbons—chains of carbon atoms with hydrogen atoms attached—and burn them in oxygen to create water, energy, and carbon dioxide.[11]

There are good reasons why fossil fuels have achieved such dominance. It's hard to imagine an energy source that is easier to use: you need only dig it out of the ground and set it on fire, no lasers or force fields required. In the past, the price was low relative to other energy sources. Fossil fuels are able to respond quickly to demand, in just a few seconds in some cases. They don't stop operating because it's overcast or there's no wind. They generate the right amount of power for anything, whether you're trimming grass with a mower or lighting a city. And even the fossil fuel power plants with the largest output barely take up any space—less than one square kilometer to power almost 3 million homes. A wind farm of equivalent power would take up substantially more land area than Washington, DC.[12]

Unfortunately, the overwhelming consensus is that we can't continue to rely on fossil fuels; eventually they'll just run out. As neatly summarized by David J. C. MacKay in *Sustainable Energy—Without the Hot Air*, the world was lucky enough to start the Industrial Revolution with 2 billion years' worth of accumulated energy reserves in the form of fossil fuels. But at current production, there are 50 years left of oil and gas, and 132 of coal. It's likely that we'll find more fossil fuels if we look hard enough, but it won't change the basic fact that they'll run out sooner or later.[13]

Even if we were to find more fossil fuels, the security and environmental problems they pose are so serious that we need to wean ourselves off them anyway. They tend to be concentrated in a few geographical regions, lending power to those countries that, more or less by chance, have them. In 1973, a group of countries cut production and exports of oil so effectively that the price rose four times over. Energy security is itself a good reason to kick the fossil fuel habit.

Burning fossil fuels is also a major source of air pollution. Air pollution is thought to be a contributing factor in the deaths of nearly twenty-nine thousand people a year in the UK and, according to World Health Organization estimates, 8.8 million people worldwide (more than smoking). Even when it's not killing you, pollution from fossil fuels has pernicious effects: one study of American students found that those who moved from a school with low ambient pollution to another with higher pollution experienced significant decreases in test scores and increases in behavioral incidents and absences. In the world's most polluted cities, for example in India and China, the negative effects are even worse.[14]

On a global scale, it's the carbon dioxide, or CO_2, produced by burning fossil fuels that's the biggest problem. Human-caused climate change is predominantly created by the greenhouse effect of CO_2; the CO_2 comes from energy generation. Levels of CO_2 in the atmosphere have risen dramatically over the last few hundred years, increasing almost 50 percent since the Industrial Revolution.

The Earth hasn't seen levels of CO_2 this high for at least eight hundred thousand years.[15]

Dr. David Kingham, vice chairman of Tokamak Energy, told me that the deep de-carbonization of the planet was going to be a bigger, more important challenge than getting to space. There's certainly more at stake.

We don't yet fully know the consequences of messing with the global climate system, but what we do know is very bad. The planet is a complex machine; each part interacts with every other part. The best models predict that human-caused climate change will create rapid changes in temperature, which will have a large and negative effect on life on Earth—human and otherwise. If we breach a 2 degree Celsius (approximately 3.6 degrees Fahrenheit) increase in average warming, a third of the planet will experience life-threatening heat waves every five years. Virtually all coral reefs will be wiped out. Crop yields will fall, and there will be extensive coastal and river flooding, huge climate-related loss of biodiversity, and an increase in extreme weather patterns. Many of these changes are irreversible, at least on human timescales. Getting it wrong on climate change could be catastrophic.[16]

Scarily, the catastrophe is already happening. The global average temperature for 2019 was 1.1–1.3 degrees Celsius (approximately 2.2 degrees Fahrenheit) above the levels of 1850–1900, according to several independently curated data sources. The World Health Organization (WHO) estimates that climate change is contributing to the deaths of around 150,000 people worldwide per year already.[17]

In 2018, the Intergovernmental Panel on Climate Change (IPCC) revealed just how little time we have left to act, injecting new urgency into the energy crisis. To avoid the worst effects, it was previously thought that the rise in temperature should be kept to, at most, 2 degrees Celsius. But as more data are gathered and more supercomputer simulations run, it seems that this just isn't going to cut it. The IPCC now recommends that we should keep warming to under 1.5 degrees Celsius (approximately 2.7 degrees Fahrenheit). We've already breached 1 degree Celsius, so this leaves very little room, and time, to act. Newspapers reported this as, "We have just twelve years to limit devastating global warming." Such a specific timeline invites skepticism; a complete CO_2 emissions halt a day after the deadline would surely still be welcome. But the point stands that we need to effect a massive reduction in greenhouse gas emissions as quickly as possible if we wish to avoid the worst climate outcomes.[18]

Perhaps you think that we're making progress on clean energy. Partly, yes. However, the consumption of oil, coal, and gas actually *increased* between 2009 and 2019 (though coal use *may* now be falling for good). Perhaps you've heard that electricity generation is greener because it's easier for renewables to displace fossil fuels for that purpose. But even in restricting the discussion just to electricity, we encounter bad news: the shares of both non-fossil fuels and coal in electricity generation in 2019 were broadly unchanged from their levels of twenty years earlier. In 2019, coal was still the single biggest source of energy for making electricity.[19]

Figure 2.1 shows how fossil fuels absolutely dominate world energy production. Fusion's cousin, nuclear fission, supplies around 4 percent. Biomass refers to biologically grown materials that can be burned for energy, including wood. Solar, wind, and other forms of renewables are barely large enough to feature except for a sizeable sliver attributable to hydroelectricity (6 percent). There is a lot of talk about solar and wind, but on the scale of the planet's energy consumption, their contributions are so tiny that you can barely make them out. The sad truth is that our energy mix means global emissions of carbon dioxide are still growing: on average, by 1 percent a year over the last decade.[20]

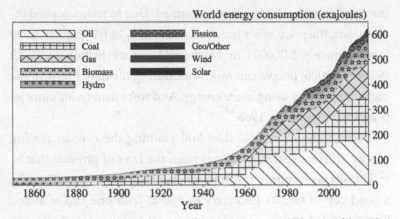

Figure 2.1: World energy consumption by generation type tells a tale of fossil fuel domination, while wind and solar are so small that you can barely see them. Biomass includes wood-burning and biofuel.[21]

How Can We Solve the Energy Crisis?

What the world needs is solutions. Star builders think they have one, but they're not the only ones.

Some people are saying that we can become radically more efficient in our use of energy. It's a nice idea, and we should certainly be doing it where we can. Improvements in technology *have* led to replacement of old style incandescent lightbulbs (which converted most of their energy into heat rather than light) by LED bulbs that are five times as energy efficient. And due to commercial pressures, jet planes became twice as fuel efficient per kilometer between 1968 and 2014.[22]

Energy savings are great, but they're extremely unlikely to offer escape from the mess. One underappreciated reason is that if products are cheaper or better, people tend to buy or use more of them. This rebound effect means that technological improvements that make products consume less energy per unit can actually *increase* the amount of absolute energy consumed. Due to technological innovations, the price of six hundred hours of light fell from £35,000 (approximately $47,000) in fourteenth-century Britain to £2.89 ($3.86) in 2006; people can now work during the dark hours of the night, but only by using more energy. And we're using a lot more jet planes than we did in 1968.[23]

Perhaps the most fatal flaw undermining the ever-increasing energy-efficiency strategy comes from the laws of physics: that is, most tasks take a minimum amount of energy to perform. To make a good cup of tea, it's necessary to boil at least one cup of water. There's a fixed and immutable energy cost to get a cup of water to 100 degrees Celsius (212 degrees Fahrenheit) that you can't reduce with efficiency savings: you can't beat physics.

There's evidence that carbon taxes would help reduce demand for the kind of energy the world needs to cut back on.[24] Economists don't usually agree on *anything*, but *most* are in favor of carbon taxes as a way to cut emissions and combat climate change.[25] But any serious solution will need to address the supply of energy too. For instance, to fill the gap that will be left if fossil fuels are removed from the mix, the size of scale-up in clean energy we will need is terrifying. The IPCC say that no more than 40 percent of our energy should come from fossil fuels by 2050. To achieve the IPCC goal will mean deploying carbon-free energy sources at a rate that is unprecedented.[26]

Star builders have been keen to tell me that massive adoption of renewable energy (solar power, wind power, tidal power, hydroelectricity, and so on) is part of the solution. Renewables have some fantastic advantages—the most obvious being that they will never run out. They also already produce net energy, unlike nuclear fusion. In recent years, the prices of some renewables have plummeted too. In particular, solar power just keeps getting cheaper.[27]

But, the star builders say, renewables alone aren't going to cut it.

They can't provide energy at the scale, growth rate, and level of consistency that is needed to power the entire planet. As Sibylle Günter at the Max Planck Institute told me, the scale required puts them in competition for space and resources that serve other human needs, like food. The scaling problem is partly attributable to another inescapable bit of physics: the energy that renewable facilities tap into is diffuse. It's impossible to mine a rich "vein" of sunlight because it's distributed over the surface of the Earth. Coal, by contrast, is a dense ball of stored sunlight. The same diffusion problem posed by sunlight applies to energy extracted from waves and the wind.

With renewables, the small amount of energy obtained per amount of land area, and their not being at 100 percent capacity all of the time, means that *huge* plants are needed. For the UK, which has a similar land area to the state of Oregon, to rely on wind alone for electricity would mean covering up to 17 percent of the country with onshore turbines, or 2.7 percent for solar photovoltaics.[*28] Offshore wind and tidal power also require impractically large areas, and hydro and geothermal power aren't available in enough places.[29] Not everyone wants to live next door to a large-scale renewable plant either.[30]

The renewable with the most promise, the one that could be scaled up the most significantly, is solar (though in some places, like Northern Europe, wind power will be more effective). Based on what has been learned from operating solar farms in the US, it would take two thousand solar power plants each the size of London to get close to providing current world energy consumption. It isn't impossible, but it's daunting and implausible. The UK government Committee on Climate Change thinks only about 60 percent of the country's electricity will come from renewables by 2050, and

---

*There are surprisingly large number ranges for renewable energy density; these are based on renewable sites that were generating in the US in 2016.

few believe that even solar power will be able to supply more than 50 percent of electricity worldwide.[31] Remember too that electricity is just one part of total energy use, and its share will need to increase substantially to beat climate change.

In a sad irony, climate change itself makes the haul of energy from renewables less certain. As the worldwide pattern of both weather and climate changes, it will create renewable energy winners and losers. The potential for solar power will likely rise in Europe and China but fall in the western USA, Saudi Arabia, and across Africa. Overall, the effects seem likely to be negative: solar cell efficiency drops by about 0.5 percent for every one-degree increase in temperature, and weather effects mean a reduction of direct sunlight by 5 percent.[32] Even if these shifting patterns create a *redistribution* of potential renewable energy rather than a reduction, the change presents a problem for policymakers.

Another problem is that renewables, including solar, are inconsistent. On a still day, wind turbines produce next to nothing. Even if solar panels were pasted across most of the landscape, they'd still be vulnerable to the weather. Solutions could include building cross-continental power transmitters or smoothing out the uneven supply of electricity using enormous batteries. Unfortunately, those big batteries don't yet exist. Tokamak Energy CEO Jonathan Carling is skeptical that batteries will *ever* do the job. "You can have a battery that extends the day a bit," he told me, "but to have a battery that turns winter into summer is too much to ask."

Dealing with intermittency is also expensive, especially when renewables get near to providing 100 percent of the electricity supply. If power is scarce, you need instant, on-demand power, either from batteries or from a neighboring state. When power is plentiful, there is a risk of over-generation—of power flooding the grid and causing physical damage. Faced with that imminent prospect, a nation or state might have to pay its neighbors to *take* power.[33]

Most studies put the plausible fraction of renewables' share of total global energy at 27 percent by 2050. Renewables are going to be a key part of the solution to the energy crisis, a very large part, but not the whole story. They're an absolutely necessary part of the solution, but they'll likely be insufficient.[34]

Of every star builder I've spoken to about whether a combination of existing technologies will be enough, First Light Fusion CEO Nick Hawker seems to have thought the most deeply about it, even commissioning a study by a third party.

"We looked at the future of energy in the 2030s and '40s," he told me. "It's instructive to ask what's *not* the problem. It's not cost. Solar and wind are already the cheapest form of energy generation. Full stop. And it's not about energy intermittency either. We can debate about how you're going to manage the last 10 percent of intermittency, but frankly if we're in that world we're doing pretty well."

So what *is* the problem in his view?

"It's the deployment and the maximum rate of deployment. The problem is scale, which is very bad news for climate change because none of the other options are very good, none of the other options are really proven—the only other one is nuclear fission."

Fission, now, that's an idea.

Nuclear fission already works. It produces next to no CO_2 compared to fossil fuels, and even less than solar cells do.[35] It's not affected by the weather or climate, and it doesn't require vast tracts of land. Powering the entire planet on fission would merely require fifteen thousand (1240-megawatt) fission plants. That's a lot of nuclear power stations, but they'd take up a tiny fraction of the land area that solar power would need.

I should declare now that I'm a fan of fission power, and not just because it involves nuclear physics—countries that heavily rely on it are managing to get off fossil fuels. France and Sweden have gone further than most in de-carbonizing their electricity generation, with 75 percent and 40 percent, respectively, of their energy supplied by fission. France also exports a lot of that sweet, clean electricity to other nations.

But I'm in the minority: an IPSOS poll, conducted after the tragic meltdown at the Fukushima nuclear plant, found that fission had the lowest public support globally of any source of power, including solar, wind, hydro, gas, and coal. Some countries are phasing out fission power altogether. "Fission faces a great acceptance problem," Sibylle Günter said.[36]

Nuclear fission certainly does have problems: radioactive waste, the possibility of dangerous events like reactor meltdowns, a finite supply of fuel, and the potential for the proliferation of nuclear weapons. I'll dig into nuclear risks of all shapes and sizes more in Chapter 8. Fission can also be expensive, although it doesn't have to be and the price varies widely depending on the context. The proposed fission reactor at Hinkley Point in the UK has a guaranteed price of £330 ($440) per gigajoule compared to £143 ($190) per gigajoule for the UK's latest offshore wind sites.[37]

Nick Hawker doesn't think there's anything inherent about fission that makes it expensive; after all, France's role as a major exporter of largely nuclear electricity suggests they're able to make it cheaply. He thinks it's the very tight regulations around fission that are to blame. "We're doing a pretty good job of showing we can't make it economical, and regulatory considerations suggest we simply can't do it," he says. "China can build nuclear fission, and South Korea can, but we can't."

Of course, those regulations may be there for good reasons. "When something does go wrong, it *really* goes wrong," Jonathan Carling says of fission, "to the point where you've got to evacuate portions of the population and for that reason there are lots of places in the world that just don't want to do that." He also brought up the proliferation risk and that not many people want a fission reactor, which could melt down, in their backyard.

Too often, renewables and nuclear fission are viewed as either/ or options. Honestly, looking at the scale of the challenge we face, we're going to need both. But we should take people's concerns and the risks seriously. Star builders are, and they don't think we can rely on fission alone to fill the gap between what energy renewables alone can provide and how much energy we'll need.

The Star Power Rescue Plan

What of the star builders' grand plan? How would building a star help save the planet? On paper, nuclear fusion combines many of the advantages of other power sources.

Ian Chapman is clear. "It's going to change the world," he told me, adding that preventing catastrophic climate change with star power is what gets him out of bed in the morning. "If I didn't believe that fusion was important for that I'd stop doing it and go do something else with my life," he said. "The imperative to tackle climate change is only escalating and I could see that back in 2000." Since that point, he says, we've "burned 80 percent more oil and gas. We need a seismic change in how we provide clean, affordable, safe, carbon-free energy, and fusion will be a big part of that, I'm convinced of it."

His colleague, and the physics coordinator of one of Culham's experiments, Dr. Fernanda Rimini, agrees. She told me that the planet would go on after a climate catastrophe, but she's trying to bring star power to Earth for the next generation, for our way of life. I ask about alternatives, but she points out that fission just isn't acceptable, and renewables aren't enough. She sees a future that

41

combines renewables and fusion. "Fusion could replace fission," she says, "but be more environmentally friendly."

Why are the star builders so in love with fusion? In short, they believe it will solve in one fell swoop many of the problems of the other power sources. The greatest problem of energy production is climate change: fusion will produce no more carbon dioxide than fission. How much carbon dioxide is that? Solar photovoltaics have life-cycle emissions that are 50 percent higher than fission and wind. Everything else, including biomass and hydro, has emissions that are at least twenty-five times that of fission and wind.[38]

Fusion also looks promising with respect to land use and scale. The star builders say that fusion plants will take up just a fraction of the space of an equivalent-power renewable plant. They'll need a land area similar to what is taken up by fission plants today. And fusion neatly solves the problem of consistency. The star builders say that, once cracked, fusion will provide a steady supply of energy come rain or shine.

Yet another argument for fusion is that it produces a lot less radioactive waste than fission, and what is produced won't last as long. Star builders say that it's impossible for a nuclear fusion reactor to melt down or undergo a runaway reaction. The amount of regular waste would also be vastly less than fossil fuels produce.

Given that the most devastating nuclear weapons are based on fission and fusion, it's rational to ask whether fusion power plants would pose even more of a risk of nuclear proliferation than nuclear fission plants. Seeking the perspective of someone who works at a site where the US's nuclear arsenal is maintained, I talked to NIF's Jeff Wisoff about this. He was clear: "Fusion is less of a nuclear weapons proliferation risk than fission." I'll revisit these claims later in the book.

And while energy security due to possible restrictions in fuel supply is a problem for oil, coal, gas, and the fissile material used in today's nuclear reactors, it's not a problem for fusion because almost every nation on Earth has direct access to it: the supply can be found in plain old seawater.

Fusion is a fuel-based power source, like fission or fossil fuels, so, in principle, it *could* run out. The fuel is a finite resource. It would be reasonable to wonder whether, if fusion is perfected, humanity could run out of fusion fuel and be right back in another energy crisis in a few years' time. But, as the star builders point out, we won't be running out of fusion fuel any time soon.

The two reactants needed for the simplest form of fusion (which requires the lowest temperatures) are the special types of hydrogen known as deuterium and tritium. Deuterium is outrageously common, with every briny bathtub of seawater containing five grams of the stuff. We know how to extract it too.[39]

The other ingredient for the simplest fusion reaction, tritium, is very mildly radioactive and, because it decays, doesn't exist in significant quantities on Earth. However, it can be made from another element that is extremely plentiful: lithium. Lithium exists in ores and in seawater. It's around fifty times more common in seawater than uranium, the raw ingredient for nuclear fission. While it's not currently extracted at scale, if any metal is economical to extract from seawater, it's lithium.[40]

Figure 2.2 shows how long different types of fuel would last if we relied on them, and only them, to provide everyone on the planet with the same energy use as the average US citizen. In the figure, the clear columns are the non-fusion fuels. The gray columns show how long the fusion fuels would last. The clear columns show that, by going all out on any one of oil, gas, coal, or existing reserves of uranium (assuming no new reserves were found), the planet would run down its energy supplies in fewer than one hundred years.

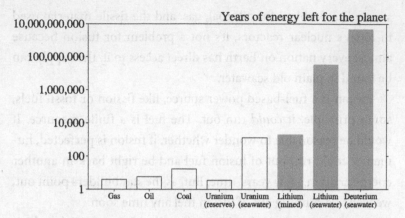

Figure 2.2: The number of years of energy left if we used just one type of fuel. Clear columns are non-fusion energy sources. The gray columns are types of fusion—deuterium-tritium fusion (using lithium from either the ground or the sea) and deuterium-deuterium fusion (which just requires seawater).[41]

Uranium extracted from seawater would enable fission to go on a lot longer.

The gray columns for fusion are labeled by the limiting factor in how long each type of fusion would last. Deuterium-tritium fusion is limited by the amount of lithium, either from ores or from seawater, because lithium must be used to create tritium. Deuterium-deuterium fusion is limited only by the amount of deuterium in the oceans. The simplest form of fusion, using mined lithium, would last two thousand years. Lithium extracted from seawater[42] would stretch the fusion energy supply to 30 million years. Deuterium-only fusion would last over a *billion years.*

Our planet's supply of fusion fuel is effectively limitless. Deuterium-tritium fusion would last long enough for continents to move, and species to rise and fall. That's a geological timescale. Energy from fusion using deuterium alone would last for *astrophysical* timescales—long enough for the Sun to exhaust its own hydrogen fuel supply and expand to swallow the Earth. Fusion is a power source that could go on until the planet itself is uninhabitable.

Fusion buys the planet time. But in another sense, time is running out. When I visited Ian Chapman at the Culham Centre for Fusion Energy, a crop of fresh PhD students just beginning their star-building careers was presenting posters on their work and discussing it over a modest buffet lunch of sandwiches and potato chips. I had some time to kill before I went upstairs to meet their boss, so I asked a group of them about their motivations for working on fusion. I was especially interested if, as people likely to inhabit the planet longer than their older peers, they were drawn to fusion because of its potential to help with climate change. One or two said yes, fusion was going to help stop climate change. But the majority were skeptical that fusion energy would be ready in time to meet the IPCC's dramatically short deadline for averting a climate catastrophe. They still believed that fusion could and should be an essential part of Earth's energy supply in the future, but they didn't believe it would arrive quickly enough to keep climate change in check. I was taken aback by their view, which was so different from that of more senior-ranking star builders. But given that the development of star machines has taken decades so far, I thought it was a good challenge, and I decided to put it directly to their boss. Upstairs, I asked Ian Chapman if fusion would be ready too late to make a difference; he shook his head emphatically.

"I don't think it's true. We don't just get to 2050 and there's a cliff edge and the world just stops or we've saved it," he told me. "We'll be on the right path, but there'll still be a whole load of carbon technology we'll need to displace."

He also said that the IPCC deadline wasn't a reason not to pursue nuclear fusion.

"You've got to remember that 80 percent of our energy is oil and gas. You can't just say by 2050 we've got to be net zero—how are you going to deliver that? Eighty percent! That's a huge challenge, and to be honest we're not going to meet it with current technology."

He's strongly pro renewables, and pro investment in renewables. But he doesn't think they'll be enough on their own. "People are totally blasé about it," he says. He is concerned, angry even. "We're not behaving like it's an emergency, and at some point that will bite."

Other star builders agree that the quicker fusion energy is delivered, the better for the climate.

"I feel that in many ways, time is our biggest enemy. If we want fusion to be a big factor in combating climate change—tick tock, tick tock," the director of MIT's Plasma Science and Fusion Center, and cofounder of fusion start-up Commonwealth Fusion Systems, Dennis Whyte has said.[43] Another star builder told me that eventually—whatever the specific timescale—the world will be driven to star power.

Star builders at start-ups were among the most bullish about the potential of fusion to defeat climate change. Scientists at both Tokamak Energy and First Light Fusion talked about building multiple net-energy-gain reactors in the 2020s or 2030s.

Given the long history of star building and the frustratingly slow progress thus far, it may seem surprising that so many star builders are so upbeat about the prospects of stopping climate change with fusion energy. In subsequent chapters, I'll explain why they're so strongly convinced they can get it working in time.

Nuclear fusion isn't the only energy source that's taken its sweet time to come to fruition, by the way. The utopian John Adolphus Etzler had the bright idea of solar power in a visionary 1833 book, and the inventor Charles Fritts's first solar cell appeared in 1883. Newspapers in 1891 jubilantly promised that "the day is not unlikely to arrive before long" when economical solar power would be driving "all the engines on the Earth." Forty years later, in 1931, they assured readers that "use of solar energy is near a solution." Close to two centuries after the idea was proposed, the cost of solar power has become competitive with fossil fuels, although installed capacity is comparatively tiny. Maybe there's hope for fusion yet.[44]

And yet . . . nuclear fusion is qualitatively different. Even Charles Fritts's first solar cell managed a net energy gain, harvesting 1 percent of the sunlight energy that fell on it. Ignoring the energy cost of creating the cell in the first place, that first cell produced energy. So far, the only widely known artificially triggered fusion reactions that have produced net energy gain have been the ones in hydrogen bombs.

Nuclear fusion has always been essential for life on Earth to exist and to flourish. Essentially, all of our energy comes directly or indirectly from the massive, fiery nuclear fusion reactor in the sky that we know as the Sun. After all, solar power, the most promising renewable, is just indirect nuclear fusion energy. The star builders want to cut out the intermediary. They recognize the difficulty of their quest and that the stakes are as high as any we as a species have faced. But to have a hope of building a star and saving the planet, they need to understand nuclear fusion's secrets.

CHAPTER 3

ENERGY FROM ATOMS

"A star is drawing on some vast reservoir of energy by means unknown to us. This reservoir can scarcely be other than the sub-atomic energy which, it is known, exists abundantly in all matter; we sometimes dream that man will one day learn to release it and use it for his service. The store is well-nigh inexhaustible, if only it could be tapped."

—Arthur Eddington, "The Internal Constitution of the Stars," 1920[1]

This book is about scientists' attempts to unlock energy from within the atom, and the star builders owe a great debt to the person who did more than anyone else to show the world this was possible.

That person is the physicist Ernest Rutherford, who, in the first decades of the twentieth century, discovered the structure of the atom, carried out the first nuclear reaction (without realizing it), and led the teams that discovered both nuclear fusion and nuclear fission reactions. But even the first ever artificial fusion experiment performed by Rutherford would show that achieving net energy gain was going to be very, very hard.

Ernest Rutherford was a brilliant, innovative, and hardworking physicist whose character and creativity won him many successes throughout his career. He helped to develop an early form of sonar;

he built a world record–beating device for detecting radio waves at a distance; and his sensational early discoveries in radioactivity won him the 1908 Nobel Prize for Chemistry. He's so important to the history of science that the UK's Royal Society has kept some bizarre personal effects of his, including a potato masher. He was generous of spirit and, also, averse to affectation. His fellow physicist Niels Bohr said, "Although Rutherford was always intensely occupied with the progress of his own work, he had the patience to listen to every young man, when he felt he had any idea, however modest, on his mind." Bohr also observed that Rutherford had "little respect for authority and could not stand what he called 'pompous talk.'"[2]

Rutherford's work on the atom began with his move to the University of Cambridge's Cavendish Laboratory, from New Zealand, in 1895. Rutherford's supervisor at the Cavendish was the great physicist Joseph John "J. J." Thomson, the discoverer of the electron: the smallest distinct particle that makes up the atom. In 1904, Thomson had put forth a theory that could explain the facts about atoms as they were then known: the "plum pudding" model, a Christmas pudding–like dessert containing raisins (rather than plums). The model posits that atoms are made up of a positively charged "pudding" embedded with negatively charged "raisins" (electrons).[3]

In 1909, Rutherford moved to the University of Manchester, where he came up with experiments that tested his former supervisor's theory. Rutherford ingeniously showed that—rather than a plum pudding—most of the atom was made up of empty space, with a concentrated lump of positively charged matter right in the middle. We now call this lump the nucleus (from which we get the word "nuclear").[4]

The basic structure of Rutherford's model is similar to how scientists think about atoms today: a massive nucleus in the center, occupying a small fraction of the space, and a cloud of very light, negatively charged electrons that are dispersed throughout the atom. Within the nucleus, there are two types of particles that have very similar masses: neutrons and positively charged protons.

The nucleus of the most common element in the universe, hydrogen, consists of just a single proton. The number of protons in an atomic nucleus determines what kind of atom it is: one for hydrogen, two for helium, three for lithium, and so on; adding one more proton to the nucleus turns it into a different element. Almost all chemistry, including reactions like burning or wine turning into vinegar (oxidation), is determined by the equal number of protons and electrons. The players in chemistry are the electrons, which are shared, swapped, or moved around in chemical reactions. As Bill Bryson observed, "Protons give an atom its identity, electrons its personality."[5]

Neutrons, which have no charge, aren't involved in chemistry, which is why atoms are usually called by one name regardless of how many neutrons are in the nucleus. For most purposes, there is no distinction between the two types of nitrogen in the Earth's atmosphere; both have seven protons, but one has seven neutrons while the other has eight neutrons. The name for atoms with the same number of protons but different numbers of neutrons is isotopes. Because both isotopes of nitrogen have the same number of electrons, their chemistry is almost exactly the same.

But for nuclear physics, what's in the nucleus matters. So the difference between isotopes is important, and can be the difference between life and death—as you'll see in a later chapter when we come to the story of Bikini Atoll and the *Lucky Dragon*. Fusion and fission are nuclear reactions because they change the number of protons and neutrons in the nucleus. Deuterium and tritium, the fuel for the star builders' preferred form of fusion, are isotopes of hydrogen. They all have one proton, but deuterium and tritium have neutrons too, which makes them more massive. The names give a clue as to how they relate to the more common form of hydrogen that has a single proton in its nucleus—deuterium is approximately twice as massive as usual hydrogen, while tritium is three times as massive. There's one neutron and one proton in deuterium; and two neutrons and one proton in tritium.

Even the broad structure of the atom that Rutherford discovered was revelatory to scientists of the day. Ernest Mach, after whom the Mach number* is named, said in 1910 that "if belief in the reality of atoms is so crucial, then I renounce the physical way of thinking, I will not be a professional physicist, and I hand back my scientific reputation."[6] The new theory was a huge paradigm shift that excited strong feelings. Rutherford clearly enjoyed mashing theories just as much as potatoes.

Rutherford is a hero for physicists because he did so much to shine a light on the structure of the atom. But he's held in special regard by star builders because of experiments that went even deeper into the nascent field of nuclear physics.

One of the ways that physicists in Rutherford's time were studying atoms and nuclei was by accelerating them to high speeds, and thus high energies, before smashing them into other atoms to see what happened. These very high energies allowed the nuclei of atoms to get close enough to interact, rather than more passively bumping off one another. A lot of knowledge about subatomic physics has been gleaned in this way, and the tactic of smashing particles for knowledge continues today in experiments conducted at the Large Hadron Collider at CERN, the European Organization for Nuclear Research.

In those early days of atomic research, the machines were a lot smaller, but the discoveries were just as big, if not bigger. Rutherford himself oversaw a series of experiments in which nuclei were accelerated to very high energies and smashed into other nuclei.

*The ratio of the speed of an object to the speed of sound in the surrounding medium, best known in the context of supersonic flight by aircraft.

In one, in 1917, Rutherford performed the first artificial nuclear reaction; he transmuted nitrogen by firing a helium nucleus into it. Rutherford didn't quite realize what he had done at the time: he thought he'd split the nitrogen apart, but in actual fact he had turned it into oxygen, releasing a hydrogen nucleus in the process.[7]

But it is two subsequent experiments that Rutherford was involved in that really started the quest for nuclear energy.

Atomic Energy

First, in 1932, two physicists working under Rutherford's guidance at the Cavendish made a huge nuclear discovery that made headlines around the world and provided the first proof that energy could be unlocked from within the atomic nucleus.

The experiment was performed by John Cockcroft and Ernest Walton. Cockcroft and Walton were accelerating protons to high energies and slamming them into lumps of lithium. Most protons didn't do anything interesting. But just occasionally—ten for every 1 billion fired—a proton would strike a lithium nucleus and split it clean in two. A nucleus of lithium-7 (the most common isotope of lithium, with three protons and four neutrons) became two helium-4 nuclei. The proton caused the lithium nucleus to have a surfeit of positive charge and so repulse itself, elongate, and break apart. It was pure alchemy; the researchers put in protons and lithium, and got out helium. They'd discovered that some atoms are fissile—that is, they can be split apart—an extraordinary insight into what is possible with the matter that makes up, well, pretty much everything.[8]

Cockroft, Walton, and Rutherford had split the lithium atoms. But the experiment was most interesting for what they saw when they added up the energies involved. They'd put 0.6 mega-electron volts (MeV) of energy into accelerating each proton—a tiny amount on a human scale, one-tenth of a million millionth of a joule. While you've been reading this sentence, you will have used up as many as one hundred joules. But at the atomic scale, giving a proton 0.6 MeV of energy is a bit like strapping a fly to a rocket. As big a number as this is, it's dwarfed by the energy that came out of the reaction. In producing the two helium nuclei, 17 MeV of energy was liberated. The colliding protons had released almost thirty times the energy that had been put into them!

Cockcroft and Walton won a Nobel Prize. Later, it was discovered that more massive isotopes could be split to yield energy in a repeating chain of nuclear fission reactions, knowledge that would ultimately lead to the atomic bomb and today's nuclear fission reactors. Cockcroft became director of the UK's Atomic Energy Research Establishment, where he arranged for the very first star machines to be housed and supported the construction of what the star builders of the day hoped would be the world's first net-energy-gain fusion device. Those hopes were dashed, but Cockcroft was nonetheless one of the most important early star builders.

But it's his second experiment that's the most important for the star builders, and the one that they can trace their entire creed back to. In 1934, Rutherford and two other physicists, Mark Oliphant and Paul Harteck, were using an improved version of the Cavendish's particle-smashing machine. They were firing so-called heavy hydrogen—what we now call deuterium, the isotope of hydrogen with one neutron in the nucleus—at other atoms of deuterium in a target. What they found was that—every so often—a reaction happened in which a deuterium nucleus with 1 MeV of energy combined with another deuterium nucleus to transform into a tritium nucleus, a proton, and 4 MeV of energy.* They'd put one unit of energy and two

---

*They also found a second reaction that produced an isotope of helium, a neutron, and 3.3 MeV of energy.

deuterium atoms into their reaction and somehow ended up with a bigger nucleus (tritium) and four times as much energy out. They'd fused the deuterium together to make a new isotope, tritium: they had discovered that fusion of atomic nuclei was possible.[9] They'd shown beyond doubt that simple, common elements, like hydrogen, could be combined to make bigger, more scarce elements. This single idea could explain how so much of the rich zoo of atoms in the universe came to be, and it would soon be used to explain how the Sun shone. It was a revelation. What's more, it put the power to re-mold creation into the hands of humanity.

Fusion is the reaction that alchemists dreamed of—it really can turn base elements into gold.[*10] It's the reaction that makes the entire universe a Lego set—with the right base blocks, you can build anything. Is it any wonder that star builders are so captivated by it?

However, fission and fusion are more than just ways of rearranging the atomic building blocks that Rutherford and his colleagues had helped discover. Some rearrangements, like the fusion reactions that make the Sun shine, the splitting of lithium, and the fusion of deuterium with itself performed by Rutherford, release energy. Not chemical energy, because they don't involve electrons, but nuclear energy, and a lot of it. What scientists had discovered was a new energy source hidden deep within the atom.

Star builders like Mark Herrmann, at NIF, and Ian Chapman, at the Culham Centre for Fusion Energy, are using deuterium-tritium fusion reactions, which produce even more energy than Rutherford's. The two isotopes of hydrogen fuse together to create a helium nucleus, a neutron, and energy. In every successful reaction, scientists put just 0.1 MeV in and get a whopping 17 MeV out; that's 170 times as much energy out as went in. On an atomic scale,

*Three scientists did this in 1941 by bombarding mercury with neutrons. But the gold they created was contaminated with radioactivity, and the experiment would have cost much more than the gold was worth. So not a path to fortune.

this is huge: 17 MeV is two million times more *per reaction* than you'd get from a single fossil fuel reaction.

But physics is about understanding, not just documenting. Even after the discovery of fusion, physicists didn't know why or how it was happening—they didn't know what combinations of atomic Lego blocks would fit together, or where the energy was coming from. If you're a star builder and you're interested in both releasing and using nuclear energy, you need to know more about *how* fusion happens.

A first clue as to what was happening came from one of J. J. Thomson's other students, a physicist called Francis Aston. In 1920, Aston came up with a more accurate and precise way to measure the masses of atoms. Surprisingly, he found that the masses of atoms weren't in exact whole number ratios, with, say, helium being four times the mass of hydrogen because it has four times as many particles in its nucleus. That's what you'd expect if you thought that protons and neutrons were like Lego blocks. Instead, the ratios were very slightly different: helium had a mass that was 99 percent of the mass of four hydrogens. This difference seems small and inconsequential, but it is the very reason why nuclear fusion works.

The physicist and great popularizer of science Arthur Eddington was struck by this apparent mistake in the arithmetic of the universe. Eddington was a nuclear visionary, suggesting long before Rutherford's fusion experiment that the power source of stars was subatomic in nature. Eddington reasoned that four hydrogen atoms should really have *exactly* the same mass as one helium atom, not 99 percent of the mass. The only explanation was that in going from a single building block (in the form of hydrogen) to four (in the form of helium) some of the mass must disappear. This was pure heresy: mass didn't just disappear. Matter might change form, say from a solid to a liquid, but the mass of the end products was always the same as the mass of the reactants. Conservation of mass was a central tenet of physics.

Eddington wondered whether a theory published by a brilliant physicist he greatly admired could shed light on the puzzle. It said that mass and energy were different sides of the same coin. Mass could become energy, and energy could become mass. It was a wild idea. Impressively, the physicist in question had published this theory along with three others that challenged fundamental concepts in physics in 1905. His name was Albert Einstein.[11]

The Secrets of Atomic Energy

The theory of Einstein's that Eddington had in mind to explain the minute differences in mass of atoms said that the relationship between an object's mass, when it is not moving, and its energy is

$$E=mc^2$$

where E is energy, c is the speed of light (a gut-wrenching 300 million meters, almost a billion feet, per second), and m is the mass difference that Aston found. With effortless parsimony, this simple equation explains what's going on in both nuclear fusion and nuclear fission.

Imagine a scale that measures the number of particles in an atomic nucleus. On one side of the scale, carbon (which has six protons and six neutrons) would measure exactly twelve units. Now put the inputs into fusion (deuterium and tritium) on one side of the scale and the outputs (a neutron and helium) onto the other.

On the side of the scale with deuterium and tritium, you'd have five particles between the two nuclei: three neutrons and two protons. But the scale would read 5.0304 units; almost 5, but, importantly, a little bit more. On the other side of the scale, the neutron and helium nucleus that are created have a mass of 5.0113 units. The scales would tip ever so slightly toward the deuterium and tritium. The difference between them and what comes out of the reaction is tiny, just 0.02 units of mass. Einstein's equation tells us that it's anything but tiny in terms of energy.

The m in Einstein's equation takes the apparently small value of 0.02 units, but it gets multiplied by a very big number, the square of c, and this means that even a tiny amount of mass converts into an outrageous amount of energy. Plugging 0.02 units of mass into $E=mc^2$ tells you that a whopping 17.6 MeV of energy was released from mass—just as found by experiment!

In fission, the same phenomenon occurs; the masses of the left-over pieces are a bit less than what went into the reaction. In both fission and fusion, Einstein's theory says how much energy will come out for a given change in mass.

With Einstein's equation, star builders are equipped to understand what combinations of isotopes will deliver energy if fused. One of the most amazing insights of these nuclear theories is that a little of the star builders' preferred fusion fuel goes a very long way. Rich fuels produce more energy per amount used than poor fuels, or, put another way, they have higher energy density. Kilogram for kilogram, burning coal releases about twice as much energy as burning wood. That's why coal was so important to the industrial revolution and, in part, why it's the most popular source of energy on the planet today. Crude oil releases slightly more energy than coal. The most high-energy-density chemical fuel is hydrogen gas, which releases eight times as much energy as wood. Burning coal is a chemical reaction; it swaps electrons around. It doesn't change nuclei. If you want to release a lot of energy, you need to go nuclear.

Deuterium-tritium fusion, the kind of fusion that most star

builders are doing, releases 10 million times the amount of energy per kilogram as coal. Ten *million*. If you had a fusion reactor in your house, you'd have to go to the deuterium-tritium shed once for every 10 million times you went to the coal shed. What this means is that the mass of a single cup of water contains the equivalent energy of 290 times what the average person in the US uses each year. The mass of an Olympic swimming pool contains an amount of energy in excess of total world annual energy use.

Even the fission reaction that powers today's nuclear plants isn't quite as energy rich kilogram for kilogram as fusion is; fission releases only 25 percent as much energy. In fact, there is only one reaction in the universe that releases more energy per kilogram than fusion, and it's the annihilation of matter and antimatter into pure energy so that no mass at all is left over. This reaction needs antimatter, which doesn't exist anywhere in the universe in large quantities (let alone on Earth). Which means that nuclear fusion is the most energy-rich fuel available to humanity.

Now it's clear why fusion fuel could provide energy to the planet for millions, perhaps billions, of years: not only is it plentiful, we don't need very much of it.

But there's a problem. Einstein's equation can tell star builders how much energy they can expect to release, if the reaction happens. But how do they know what reactions can happen? Are some allowed and some not?

Fortunately, the star builders can rely on decades of work that have gone into understanding nuclear reactions. At the root of what can and cannot happen are the fundamental forces of nature. We'll be seeing them again and again in this book, not least because they dictate how everything in the universe happens. And I do mean *everything*. They lie behind phenomena as diverse as the chemical reactions in a car engine, the creation of the elements, the way a stone falls, and the ability to see. It is these forces that allow objects to interact with each other; if what you're sitting on now did not hold you up with a force, then you'd simply pass through it like a ghost.

The four forces are gravity, the electromagnetic force, the weak force, and the strong force. They are what make sense of the abstract properties that particles have, like charge and mass. Electrical charge determines how strongly a particle interacts through light and electromagnetism; mass determines how strongly a particle feels gravity.

The most familiar force is gravity because it keeps you from floating off into space. Electromagnetism describes electricity and magnetism—you can see how the two are related by placing a compass next to the cord of a hair dryer or electric kettle with the power on; the needle stirs due to the magnetic field created by electrons moving in the power cord. The weak force is responsible for radioactive decay. The strong force is the glue that holds together most of an atom's mass in the nucleus.

The forces have different strengths. The strong force is the strongest, then the electromagnetic force, then the weak force, and finally gravity. It might be surprising that gravity is the weakest of the four forces, especially as astronauts have to use rockets to escape the Earth's gravity. But even a puny bar magnet can overcome the gravitational attraction of the entire planet when you use it to suspend a paper clip above the ground.

The forces act over different ranges too. Gravity has infinite reach and is solely attractive. It doesn't get canceled out by an "antigravity." These unusual properties are why it's gravity that determines how the structure of the universe evolves, despite its relative weakness. Electromagnetism also has infinite range but comes in positive and negative versions that tend to cancel out on large scales.

In contrast, the strong force has a very short range, and it's this short range that determines how big atomic nuclei can get. It is so strong that it can bind protons together in nuclei even though they repulse each other due to their same electromagnetic charges. When the strong force is acting between particles within an atomic nucleus, it has a different name: the nuclear force.

The nuclear force says that for energy to be released by fusion, two slightly more unstable atoms—that have proportionately less glue-like nuclear force—must be combined to create a more stable nucleus. For energy to be released from fission, a big, ponderous, and unstable atom must split into more stable smaller atoms. The common atom that is the most stable of all, the Goldilocks of nuclear physics, is the isotope of iron with fifty-six particles in the nucleus, or Fe-56. Atoms smaller than this tend to release energy if fused together (like hydrogen and its isotopes) because the reaction makes bigger atoms that are closer to Fe-56. Similarly, atoms bigger than Fe-56 tend to release energy when split.

The forces determine which atoms we need to unlock the most energetic reaction we can ever hope to use as a species. Great, let's do some nuclear physics!

The Problem

Not so fast. There's a catch, a big one, and it's why getting *net* energy from nuclear reactions is hard.

Remember Cockcroft and Walton's nuclear reaction that produced twenty times the energy they put in? It only succeeded ten times for every 1 billion protons that they accelerated. To produce net energy gain from the experiment as a whole, they'd have needed it to succeed more than one time in every thirty. Their experiment actually used substantially more energy than it produced. The fusion reaction that Rutherford first demonstrated was similar, with a chance of one in a million—so low that the machine used much more energy than it produced.[12]

Doing nuclear physics the particle smashing way is like throwing darts blindfolded: you might know the rough direction, but you're not going to get many bull's-eyes. Einstein himself said that the "likelihood of transforming matter into energy is something akin to shooting birds in the dark in a country where there are only a few birds." Rutherford agreed, saying of one nuclear experiment that "We might in these processes obtain very much more energy than the proton supplied, but on the average we could not expect to obtain energy in this way. It was a very poor and inefficient way of producing energy, and anyone who looked for a source of power in the transformation of the atoms was talking moonshine."[13]

Just like throwing darts at random, there are probabilities for "hitting" the target when you're firing one particle at another and hoping that you'll get a nuclear fusion reaction to happen. And just like when I throw darts, the probability of a hit is low.

The reason is that the electromagnetic force repulsion between the protons of the two nuclei kicks in much sooner than the nuclear force that snaps them together. From the point of view of the accelerated particle, it's like climbing a steep hill (electromagnetic repulsion) with a deep valley (nuclear force attraction) on the other side. Get the approach just right, and the incoming particle will have enough momentum to make it over the top of the hill and into the deep valley on the other side: fusion can happen. But most of the time, the incoming particle doesn't make it—even when it has the perfect amount of energy, it's still much more likely that the two nuclei will glance off each other rather than fuse together.

The favorite fusion reaction of star builders is the one between deuterium and tritium, which, compared to other fusion reactions, is easy to do. It's one hundred times more likely to result in fusion than the deuterium-deuterium reaction, for the same input energy. Deuterium-tritium fusion also releases a lot of energy when successful, which is why NIF, the Culham Centre for Fusion Energy,

Tokamak Energy, and First Light Fusion are all planning to use it to deliver net energy gain. There are only a handful of star builders not using this reaction.

However, even though the deuterium-tritium fusion reaction has the best odds of any fusion reaction, the chances of a successful collision are low; it's *still* throwing darts blindfolded.

The particle-smashing approach leads to many more misses than hits, and always uses up more energy than is created by successful reactions.[14] For the star builders, it isn't good enough, says Professor Sibylle Günter of the Max Planck Institute for Plasma Physics. "If we just smash individual nuclei into each other, it can be easily demonstrated that the energy required for accelerating the particles would exceed the energy gained from fusion." Particle smashing is never going to be a route to fusion energy. "The key issue is not to demonstrate the nuclear reactions," she explained, "but to gain energy in a profitable way."

For nuclear fission, physicists eventually realized that they could sidestep the problem entirely by chaining reactions together. In fact, it was Rutherford's "moonshine" comment that prompted another scientist to work out how to do it. With the right nucleus, a neutron can cause a fission reaction that releases neutrons, those neutrons can split more atoms that release neutrons, and so on, with each step in the chain releasing nuclear energy. If there's more than one neutron released, the chain grows too, increasing the energy released in each generation. Because there are stray neutrons in the environment, there's not even any need of a particle accelerator; put enough fissile material in one place (a critical mass) and a chain-reaction spontaneously begins. The detritus from so many atoms splitting is what creates radioactive waste.

Fusion doesn't split atoms, it combines them, so it can't use the same trick. For fusion reactions to come anywhere close to being able to produce net energy gain, nuclei need to collide in phenomenally huge numbers repeatedly. It's like buying millions of lottery tickets instead of one. Fortunately, nature has provided the perfect crucible for fusion reactions.

What Every Star Machine Needs

Star builders need a way to crash particles together not just once, but over and over again. To achieve this, they've turned to a strange state of matter; one that is rare on Earth, but prevalent in the rest of the universe: plasma. It's the stuff that stars are made of, and it's perfect for fusion.

Plasmas are the fourth state of matter, after solids, liquids, and gases. You've probably heard of solids, liquids, and gases, but you might not have heard of plasmas. They're very different from their cousins, in ways that are beguiling, mysterious, and—for the star builders—often frustrating. To us, plasmas are the most rarely encountered of the four states of matter. But when we turn our telescopes to the visible universe, they're the most common state of matter; they make up 99 percent of it. You can see plasmas on Earth too, whether in the form of lightning, the aurorae at the Earth's poles (also known as the Northern and Southern Lights), or fluorescent bulbs—you might even be reading this book by light generated from a plasma.

While the other three states of matter are composed of atoms, plasmas are composed of atoms that have separated into their constituent nuclei and electrons. Plasmas are clouds of ripped apart atoms that don't behave anything like the other three states of matter.

The state of matter of any element varies depending on the conditions it's in, chiefly the temperature and the pressure. On Earth's surface, nitrogen (which makes up 78 percent of air) is a gas, water is mostly a liquid, and silver is a solid. However, take nitrogen down to below -210 degrees Celsius (-346 degrees Fahrenheit), and it becomes a solid; water turns into a gas at 100 degrees Celsius (212 degrees Fahrenhiet); and you have to heat silver to 962 degrees Celsius (1,764 degrees Fahrenheit) before it turns into a liquid. Changes in temperature and pressure cause the same atoms to appear in a different form.

Temperature is a measure of average speed at the atomic scale, so more heat means more movement, and more vibration. The molecules in liquid water are bound together by relatively weak forces. Put in a little heat, and the water molecules vibrate more. With enough heat, the energy of the vibrations is more than the energy of the bonds and the molecules break free, turning into a gas. Once free, they zip along until they bounce off another molecule or a wall.

Add more energy to a gas, raise the temperature even higher, and eventually the bonds *within* the atoms will also sever—this is a plasma. Each atom of hydrogen separates out to become a nucleus and an electron. Because the electromagnetic force acts over long distances, the cloud of electrons and nuclei that exists in a plasma moves in complex synchronized motions; an endless, frenetic dance.

Only when the plasma is hot enough do nuclei collide with each other with enough energy to overcome the electromagnetic repulsion of their positive charges, and for fusion to occur. For nuclear fusion reactions to be successful, particles have to slam together at around 3 million meters (a little under 10 million feet) per second, and do so over and over again. In a successful collision, the nuclei snap together in a fusion reaction. Star builders must put lots of energy into their plasmas, obtaining temperatures of many millions of degrees, to maximize the chances that enough collisions will occur with enough energy to result in fusion. Plasmas are as important for fusion reactors as ice is for making an igloo.

While plasmas are the best bet for getting significant numbers of fusion reactions to happen, that doesn't mean it's particularly easy. Understanding and controlling plasmas is one of the biggest challenges facing star builders. The problem is that plasmas are incredibly complex. The frenetic dance of charged particles means that every part of the plasma is pushing or pulling on all of the other parts. Particle physicists, like those working at CERN, try to understand what happens when two particles collide. The long-range electromagnetic forces in a plasma mean that star builders are trying to understand what happens when 100 to 10^{15} (that's 10 followed by 15 zeroes) particles collide simultaneously with one another. Plasmas are connected, kind of like a jelly: if you push on one bit, you'll find everything else moving. The way that a plasma interacts with itself results in weird and unexpected behaviors; for example, plasma often caused a communications blackout during space shuttle reentry, and the wrapper of plasma in the Earth's atmosphere, called the ionosphere, allows shortwave radio enthusiasts to speak to one another at distances of thousands of miles.

To make progress toward understanding—let alone controlling—plasmas, physicists have had to patch together classical mechanics, quantum mechanics, laser science, nuclear physics, extreme engineering, statistical mechanics, thermodynamics, experimental physics, computer science, and electromagnetism, to name a few. The most recent addition to this mix is quantum electrodynamics, the field of physics that describes how charged particles and light interact. The difficulty of understanding plasmas is why Sibylle Günter's entire institute is dedicated to the study of them, and why the most prestigious prize in mathematics, the Fields Medal, was awarded for painstaking steps toward a better understanding of them in 2010. "It is only the plasma itself which does not 'understand' how beautiful the theories are and absolutely refuses to obey them," Hannes Alfvén said in his 1970 Nobel Prize lecture.[15] Plasma physics is complicated.

This matters. Star builders' lack of understanding of plasmas is a threat to how quickly they can reach net energy gain. It's been a problem right from the start. "Of plasma instabilities," the person who first came up with using lasers for fusion, John Nuckolls, has reflected, "we knew nothing (and had a lot to learn)."[16] And it's still a problem today. Mark Herrmann, NIF's director, confessed that he doesn't think anyone fully understands the plasma physics involved in his machine. "There are parts of the problem that we understand better and there are parts of the problem that we understand worse." Mark said that by driving so much energy into the plasma, as they do at NIF, it sets up waves within it, like a big rock chucked into a pond.

"Those [plasma] waves can conspire in lots of different ways; some of them can conspire to reflect light back out." That's not good for the success of their fusion machine.

The kind of fusion with which Ian Chapman's machine is leading the world is also highly dependent on plasma physics. The plasma created at Culham can go from flowing smoothly to chaotically, as if a stream suddenly gushed like a waterfall. One of the Centre's academic consultants, Professor Howard Wilson from the University of York, said: "I would contest that if it wasn't for plasma turbulence we would have a fusion reactor working now; we probably would have had a fusion reactor working for some decades!"

At Tokamak Energy, "plasma!" was the response I got when I asked about their biggest scientific challenge. Their competitor, First Light Fusion, said that being able to fully predict what went on in a fusion plasma was "the dream" but "a bit far-fetched." Although she is scientific director of the Max Planck Institute for *Plasma Physics*, Sibylle Günter's response when I asked if she and her colleagues fully understand the plasma physics on their fusion machine was "No, of course not." I asked Mark Herrmann whether a perfect understanding of plasma physics would allow net energy gain to be achieved. "That understanding would certainly tremendously speed up the rate of progress," he replied with enthusiasm. Plasmas are key to building a star.

Getting the plasma hot is not the only thing that the star builders need to worry about. They need to *keep* the plasma hot too; to stop its energy from escaping by putting it into a container. But these temperatures are literally too hot to handle. There is no known material that can do the job. The highest melting point of any metal is that of tungsten, at over three thousand degrees. Fusion plasmas reach over *100 million* degrees. If a fusion plasma touches a material, that material is vaporized. Catastrophically for fusion, the plasma also loses its energy and cools down.

Finally, getting more energy from fusion reactions than is put in is all about collisions between nuclei within the plasma. How close together the particles are in the plasma is just as important as the temperature. If you're in a mostly empty train carriage, you're unlikely to bump into other passengers despite the shaking of the train as it goes over the tracks. Now imagine it's rush hour. There's barely a patch of free space. No matter how small the bump, you're so close to your neighbors that you're constantly colliding with them. That's why social distancing is used to slow the spread of pandemics. And it's also why packing more nuclei into the same amount of space—or increasing the plasma density—means more collisions, and more chances for fusion.

For fusion to work, the plasma has to be kept hot, as dense as possible, *and* well confined.

Rutherford died in 1937, just one year before physicists demonstrated that chains of fission reactions could be strung together to scale up indefinitely the energy released. Because of that discovery, we've had fission power stations, and fission-based nuclear weapons, for decades. Star builders know full well what they need to show that nuclear fusion as a power source isn't moonshine either—they just haven't managed it yet.

The trio of temperature, density, and confinement are the most important properties of the plasma in any star machine. Get them right, and the odds for fusion can tip the scales in favor of net energy gain. Different star machines approach the task differently: some are hotter, some denser, and they vary in how they confine the plasma. But there are just three serious methods for confining hot, dense plasma for fusion reactions—by using gravity, magnets, or inertia. Okay, technically there is *one* more: creating a universe from scratch.

CHAPTER 4

HOW THE UNIVERSE BUILDS STARS

"I am aware that many critics consider the conditions in the stars not sufficiently extreme enough to bring about the transmutation—the stars are not hot enough. The critics lay themselves open to an obvious retort; we tell them to go and find a hotter place."

—Arthur Eddington, 1927[1]

Nature is good at fusion. Really, really good. It's galling for the star builders. On Earth, they're trying to build the most sophisticated machines ever devised to trap the stuff of stars, plasma, and initiate fusion reactions in it. When star builders explain what they're doing, their ideas can sound crazy, like barely plausible science fiction. But, out there in the universe, fusion is happening all of the time, and on scales that make fusion on our planet seem like a whisper in a whirlwind.

The star builders are inspired by the universe's bounty of fusion-produced energy. It tells them how they might achieve fusion on Earth. More important, what is going on in the cosmos shows them that fusion is possible, likely, and even ubiquitous. A glance at the night sky demonstrates that nuclear fusion is the universe's most visible, most prevalent energy source.

So how, and where, does nature do fusion so easily? And what

71

can the star builders learn from it? To find out we have to go on a journey through space and time to the very start of the universe.

If you trace everything that has ever happened back far enough, every person, every atom, every packet of light, they all lead back to the Big Bang approximately 14 billion years ago. Currently, the universe is expanding, and like dots on a balloon as it inflates, everything is growing farther apart from everything else.

Run the events backward to right after the Big Bang, and the universe was very hot, and very dense. Initially, in the first millionths of a second, it was too hot for even subatomic particles to form; there was too much energy around. As the universe expanded, it cooled, and the balance shifted so that protons and neutrons could form. The universe's regular isotope of hydrogen, with one proton, was born and remains to this day the most common element.

With what is astonishing precision, we can say that one hundred seconds after the Big Bang, nuclear fusion reactions began in earnest for the first time. The hot, dense environment hit a Goldilocks zone for fusion. Protons and neutrons fused to form deuterium. Deuterium and neutrons fused to make tritium. Tritium and protons fused to make much of the universe's helium. Energy from fusion was confined not by a reactor vessel, but by the universe itself. The chain of different fusion reactions proceeded, making more and more massive nuclei as it went.

It wasn't long, around nine hundred seconds, before the universe had expanded and cooled so much that fusion stopped. Particles had less energy when colliding; lower density meant they were less likely to bump into each other in the first place. In enough time to oven-cook a frozen pizza, the universe's first fusion factory created the four elements up to beryllium, with trace amounts of more massive nuclei.

The question is: Can those trying to do fusion on Earth learn anything from nature's first fusion symphony? Big Bang Nucleosynthesis, as this type of fusion is known, is hard to do if you don't have a fresh universe at hand, and the star builders don't. We can't learn much from it about how to do fusion on Earth; all we can say is that it's another example of how fusion is a natural process that, with the right circumstances, will happen. But there's another type of fusion reactor that's ubiquitous across the universe and by which generations of star builders *have* been inspired: the stars.

The First Fusion Reactors

When you look up at the night sky and see those pinpricks of light, you're seeing the energy released by vast numbers of enormous fusion reactors whose power output greatly exceeds the wildest dreams of Earth-born attempts to control and use fusion.

To find out how the first stars came to be, I've come to talk to someone who studies the period when they formed: Dr. Emma Chapman, a Royal Society Dorothy Hodgkin Fellow at Imperial College London. We're talking in her office on the top floor of the physics department, where all of Imperial's astrophysicists are located—presumably, to be closer to the skies they study. The room is decorated in a style that might be called "academic—classic," with books taking on a structural role. Emma is engaging and humble despite her vast knowledge, and frequently distracted by her own excitement. She's perhaps best known publicly for her battles to improve how women are treated in physics. Unfortunately, she has personal experience with the worst of this; she was sexually harassed by a senior colleague at University College London. She has used her own negative experience to improve, and prevent, the same poor treatment of other scientists by campaigning hard for universities to recognize the problem and to treat victims more fairly. Today, however, I'm here to find out about her research.

Although the stars seem like an immutable fixture of the night sky, they haven't always been around. Before there were stars, there was darkness. Four hundred thousand years after the Big Bang the universe had cooled so much that the hydrogen and helium plasma that then made up most of the visible matter began to combine to form neutral atoms and molecules. While plasmas constantly interact with, and emit, light, neutral atoms do so much less readily, and as a consequence, everything went utterly dark. It was as if someone pulled the plug on the entire universe. There was no visible light—not that anyone was there to see it, of course. Scientists call it the Dark Ages. The universe remained dark for 100 million to 400 million years—no one knows just how long.

"I study roughly the time from a few minutes after the Big Bang to about a billion years after the Big Bang," Emma tells me, which is a period that, to say the least, puts scientists' deductive skills to the test. She explains that we know a lot about planets, about our galaxy, and even about the moments right after the Big Bang. But there's a gaping hole in our knowledge of the early universe.

"We've got one point of data at about four hundred thousand years after the Big Bang," she says. "After that we've got nothing for about a billion years. If you compare that to a human lifetime, that's the equivalent to missing everything from just after the point of conception to when the child first goes to school."

It was during this period that an amazing event took place that would change the universe forever: the first stars formed and began to shine.

So how did nature build the stars from the materials available in the early universe? Gravity is the extra ingredient that turns a sprawling cloud of gas into a star. That's how it happened then, and it's how it happens today in patches of space that are stellar nurseries, like the Eagle Nebula. Because of gravity, a small cloud of gas attracts more gas to it, and that attracts yet more, clumping together. The only reason that gas clouds don't collapse under gravity is because they're supported by their own gas pressure, the kind of resistance you feel when you pump up a bicycle tire. Once a gas cloud gets big enough, the gas pressure is overwhelmed by gravity, and the matter collapses.

There's something extra special about the first stars that Emma studies. Compressing a hot gas is harder than compressing a cold one. You may have heard that hot gases like steam can push hard enough to drive a turbine, so it's no surprise that it takes more work to compress hot gases. The early universe was mostly composed of hydrogen and helium, which are less good at radiating away energy as light than other elements. As a result, they tended to form hotter gas clouds that were more resilient to gravitational collapse. Only when the gas clouds reached enormous sizes did gravity overcome gas pressure, triggering collapse.

"What you get," Emma tells me, "are these giant gas clouds that condense into a star, [which results in] stars that have about one hundred times the mass of our Sun forming. Some simulations have stars with one thousand times the mass of the Sun forming, which is absolutely huge." Those huge stars are gone now, but star formation today is still a dramatic process; in the interstellar medium, the collapses that occur are akin to taking an object the width of Australia and compressing it into something the size of a ball bearing.[2]

It was around 4.6 billion years ago that our own Sun formed from a cloud of interstellar dust, sucking up 99.9 percent of the mass of the solar system as it did so. The energy provided by gravitational collapse of that dust went into heating and compressing the matter into a dense ball. As the temperature of the core rose beyond two thousand degrees, its atoms began to be ripped apart and transformed into plasma. Over a few tens of millions of years, the temperature and density of the plasma rose, until particles began to crash together with enough speed for nuclear fusion reactions to begin: light, energy, heat—all unleashed as the solar system came alive.

Likewise, the universe's Dark Ages ended as the tender young universe was flooded with light from fusion. That light swept through the clouds of molecular hydrogen, stripping the electrons from the nuclei, and creating a hydrogen plasma in the interstellar space. Before becoming plasma, the hydrogen emitted a telltale signal that still bounces around the universe today, faint but detectable. Gaps in this signal herald the first stars and it is these that Emma Chapman and her colleagues are trying to detect.[3]

Surprisingly, Emma does her scientific detective work by listening to the radio. It's not a regular radio but a radio telescope—it can listen in on radio waves from deep space. The technology isn't as sophisticated as you might think: "A radio telescope looks like an antenna that you used to have on your car, stuck in a field," Emma says. She explains that although, ideally, astronomers would build a huge single telescope, it's impossible to build one big enough. Instead, she's part of a consortium putting up 130,000 antennas, each of which looks like a metal Christmas tree, in an isolated part of Western Australia. The trick is that they can all be connected together to act like a single giant telescope. "What we've done is built a telescope the size of Western Europe," she says. It's a big telescope, but it does more than just enable Emma and her colleagues to see extremely distant objects in space.

"With astronomy, the further you look, [the further you go] back in time," she says. This is why we see the Sun as it was eight minutes ago; it takes that long for the light to reach us. The farther objects are away, the longer it takes for the light to reach us and—in effect—the farther back in time we see. When we look at Andromeda, a neighboring galaxy that's on a collision course with the Milky Way, we see it as it was 2.5 million years ago.

What this means is that the farther away an object is through a telescope, the farther back in the past it is. Look far enough, and you can see right back to the start of the universe. Astronomers use the language of "seeing" with radio telescopes loosely: they don't mean that they can literally *see* what happened. They simply mean that they can capture and measure the faint radio signals from billions of years ago.

The humungous new radio telescope that Emma and her colleagues are using is called the Square Kilometre Array. The physical scale is not the only challenge: raw data arrive at a terrifying 157 terabytes per second. "There are no hard disks that can store that amount of data quickly enough without exploding!" Emma says. And the signal that they're looking for is just a small part of that data, a signal that's one-ten-thousandth the strength of many other radio waves they pick up. These include radio waves generated by human activity as well as the noises of our own galaxy. If you want to hear the history of the universe, you better listen closely.

With this forest of radio antennas, Emma and her colleagues hope to better sift out the telltale signals from the period when the Dark Ages ended and stars and planets were formed from clouds of neutral hydrogen gas. "Planets and stars are formed from the same cloud, from the same stuff," she tells me. "One is fusing, one is not, and that's the difference."

Since Emma is not a star builder but an early universe astronomer, I'm interested in what she thinks of nuclear fusion—though she's quick to point out that it's not her area of expertise. "If there's one thing I would put my money in, it would be that, it really would, it would be nuclear fusion."

Let the Sun Shine

Our closest star, the Sun, is both extraordinary and completely unremarkable. It's unremarkable because there are more stars like it just in our local galaxy than there are people on the planet. It's extraordinary because of its role in the human story. It's a ball of plasma 109 times the radius of the Earth and 330,000 times Earth's mass. So incomprehensible is the Sun's power and scale that many civilizations have worshiped it as a god. Almost all life on Earth has been, in some way, dependent on energy from the Sun. Given this, those civilizations don't seem too far off the mark.

The star builder who has come, *literally*, closest to our solar system's only net-energy-gain-producing fusion reactor is Dr. Jeff Wisoff, NIF's principal associate. He's been to space four times. During those trips, he conducted three space walks, some to test the tools that would eventually be used to repair the Hubble Space Telescope. Remarkably, he's not even the most experienced astronaut at Lawrence Livermore National Laboratory: he tells me that Dr. Tammy Jernigan, until recently a deputy associate director on the weapons program, has been up one more time than he has (Tammy is also his wife).

On a 4.9-million-mile journey into space and back in 2000, Jeff did one space walk in which he tested a jet pack. Remaining tethered to the space shuttle at all times, his jet pack squeezed nitrogen gas out in one direction, which had the effect of gently propelling him fifteen meters (approximately fifty feet) out into space's vacuum in the opposite direction. In orbit far above the Earth, the days are much shorter, and astronauts may see sixteen sunrises for each Earth day. Out there in space, as he fast-forwarded through a day every ninety minutes, Jeff Wisoff was momentarily closer to the Sun than any human being alive.

"One of the perspectives you get from being an astronaut," Jeff tells me, "is that the whole universe is powered by fusion energy; it's the quintessential source of energy, and when you look out and see all those stars you think 'wow' . . . if humankind can harness that in the laboratory, it's as big a deal as the invention of flight, landing on the moon, the invention of steel; it will be one of those landmark points in human history."

Like all star builders, Jeff Wisoff believes that the best way to achieve net energy gain is to re-create some of the conditions in stars that make them so good at fusion. NIF tries to create conditions that are similar to the cores of stars. Doing this with a machine is a tall order, and even Jeff describes NIF as being "more complex than the Space Station"—but, he adds, the commute is a lot easier than in his old job.

Despite the challenges, Jeff Wisoff and the others we've met are inspired by stars because they seem to effortlessly produce net energy from fusion reactions. But stars don't primarily use the deuterium and tritium fusion that he and the others are using.

The Sun is powered by a series of other fusion reactions that fit together like the pieces of a jigsaw puzzle. The raw material is the nucleus of the most common isotope of hydrogen, also called a proton. There are three steps: In the first, the protons in the stellar plasma fuse to form a single deuterium.[*] Then, the deuterium nucleus fuses with another proton to form the more rare helium-3 nucleus (two protons and a neutron). Two of these fuse together to form a regular helium nucleus, and two protons. Start to finish, the cycle fuses protons to create helium, releasing energy as it goes.

In stars, these fusion reactions happen at scales so vast, so beyond the scale of individual atoms, that it's scarcely possible to imagine them. From 600 billion kilograms of hydrogen reacting, the Sun yields 4 billion kilograms' worth of pure energy. Every. Single. Second.

There's a second chain of reactions that powers stars larger than the Sun, called the carbon-nitrogen-oxygen (CNO) cycle. It uses a carbon nucleus and four protons, and turns them into another carbon nucleus and a helium nucleus. Because these chains of fusion reactions result in new nuclei, the process is called stellar nucleosynthesis.[4]

It doesn't matter that the Sun is doing a different set of fusion

[*]There are other proton-proton fusion chains active in the Sun, but this is the dominant one.

reactions from the one utilized by scientists on Earth. Star builders can still learn from what makes the Sun's inner core, up to 25 percent of its radius, an effective fusion reactor. What this part of the Sun has is a high temperature, a high density, and strong confinement of particles and energy. So how is it achieved? The Sun and other stars have two tricks that help them with all three of these properties simultaneously: gravity and scale.

Gravity doesn't just cause stars to form, it also keeps their mass compressed to a high density—so much that a single teacup of the plasma in the Sun's core has a mass of twenty kilograms (approximately forty-four pounds). When fusion releases energy, a virtuous feedback loop is initiated—higher temperatures, more fusion; more fusion, higher temperatures. What this means in practice is that all fusion reactions in stars have an extremely sensitive dependence on temperature. For example, in bigger stars, the rate at which fusion reactions occur in the CNO cycle is proportional to the twentieth power of the plasma temperature. This means that for every doubling of temperature, the reaction rate increases by a whopping factor of 1 million. Why doesn't the energy escape? For particles with mass, gravity also helps with confinement. The gravitational pull of the Sun is twenty-eight times that of the Earth, which helps keep all that hot plasma in place. Energy from fusion also ends up as light; in fact, it's because of light from fusion reactions in the Sun that, as I write this, I can see everything in my garden. Light must be escaping the Sun eventually then, but it doesn't do so easily, because of the Sun's vast scale. Physicists often think of light in terms of packets of light energy called photons. The Sun is so dense that even photons can only make it millimeters to centimeters before they collide with charged particles in the plasma. If the Sun were a small ball of plasma, these photons would escape quickly. The Sun has a radius of 700,000 kilometers (approximately 435,000 miles). Photons are fungible—one is indistinguishable from another—so you can't track them individually. But if you could, it would take the average photon *hundreds of thousands of years* to make it to the surface.

Being so big has other advantages. Earth-bound fusion reactors are plagued by badly behaving plasma that goes unstable at the drop of a hat, ruining confinement. Stars, however, have a built-in stability system that keeps them more or less the same size. Add more mass to a star, and gravity contracts it a bit, making it more dense, and prompting fusion reactions to occur at a higher rate. Because stars have excellent plasma confinement, the extra energy raises the temperature. The result is that the star pushes harder against the force of gravity, expands, and lowers its density again. Conversely, take a little mass away and gravity binds the star less tightly, the star expands, the density falls, and fusion happens at a slower rate. But this means that the temperature drops, and the star pushes back on gravity less firmly, so it contracts again. It's a stable, self-correcting system that, like a pendulum, will return to its starting point if disturbed.

Of course, stars aren't *completely* stable—this is plasma we're dealing with, after all. Solar prominences, vast arms of plasma on the scale of the largest planets, frequently reach out of the Sun's surface and into space. The most dramatic examples are coronal mass ejections (the corona is the outer layer of plasma of the Sun). Like the Earth, the Sun has magnetic fields that emanate out from the poles like so many strands of spaghetti. The Sun rotates around 11 percent faster at the equator, with a period of twenty-five days, compared to the mid-latitudes between the poles and the equator.[5] This differential rotation stirs up the plasma, which gets caught up in the magnetic fields. Occasionally a strand of the magnetic spaghetti will develop a kink that extends beyond the surface of the Sun and breaks apart, flicking an arc of plasma into space.

Coronal mass ejections can throw plasma right at us at speeds of up to three thousand kilometers (approximately 1,864 miles) per second. Once it hits the Earth's magnetic field, it's guided to the poles, creating the Northern or Southern Lights. But when there are *huge* coronal mass ejections, the Earth's magnetic field becomes heavily distorted and plasma rains down much closer to the equator, even as far as Cuba. Such ejections can cause serious damage to power grids.[6]

Ironically, the immense scale of the Sun permits it to be, in some ways, a lousy fusion reactor. The number of fusion reactions in a chunk of Sun plasma is 3 million times fewer than what is needed, or practical, for the star builders' Earth-bound machines. A whole cubic meter (approximately thirty-five cubic feet) of stellar matter produces just 0.03 percent of the energy, per second, that an electric kettle uses. Fusion reactors on Earth need to be much more efficient with their space.

It's because of gravity and their vast scales that stars have the three key elements that star builders' machines need: temperature, density, and confinement. So could star builders ever hope to create significant net-energy-gain fusion using the exact same tricks?

The answer is: no, because the scales needed are so vast. The exact ways that stars pull off their fusion tricks are impossible to replicate in miniature on Earth. Without gravity compressing everything, it's hard to re-create the same densities as in the Sun for all but the briefest of moments. And a plasma that's just millimeters or meters in size is easy for photons to get out of, and tends to be very unstable.

The only way to mimic the Sun's way of doing fusion would be to use a star that already exists. This was the idea of the physicist Freeman Dyson, who died in 2020. He wrote about how civilizations—not necessarily our own—use energy, in an article published in 1960 called "Search for Artificial Stellar Sources of Infrared Radiation."[7] There, he began his argument with the idea that, at each phase of its development, a civilization will use more and more energy. Dyson imagined intelligent extraterrestrial species meeting their energy needs not by building a fusion reactor on their home planet but by enveloping their home star in a shell that absorbed every single joule of energy: the ultimate, full-scale fusion reactor. As Dyson himself put it, "Within a few thousand years of its entering the stage of industrial development, any intelligent species should be found occupying an artificial biosphere which completely surrounds its parent star." These constructs are now known as Dyson spheres. The point of Dyson's article was actually to propose how we might detect intelligent life elsewhere in the universe—e.g., a star that's surrounded by an artificial sphere would create a telltale signature of infrared light that telescopes could search for.

Creating a Dyson sphere is implausible, so much so that even the wildest star builders aren't contemplating it. There's no chance, for now at least, of humanity being able to directly use a star as a fusion reactor (though we can benefit *indirectly* from the Sun's fusion, using solar panels on Earth). So the scientists looking to save the planet with fusion need to think of other ways to achieve the conditions within stars.

Death Star

There's one more way for fusion to happen in space, although it's not one that any star builder would want to replicate on Earth, even if they could. Even natural fusion reactors can't go on forever, because they eventually run out of fuel. What happens next depends on the star.

The largest stars, super giants, have radii a *thousand times* that of the Sun. The hottest have surface temperatures thirty times that of the Sun. Some are so bright that you'd have to orbit them at five times the distance between the Sun and Pluto to get the same amount of light as we do from the Sun on Earth.[8]

In stars like the Sun, the cycle that turns hydrogen into helium eventually leads to the core being dominated by helium as the hydrogen runs low. However, the fusion reactions that use helium as an input, rather than an output, need much hotter temperatures. So the helium plasma sits around not doing much. In low-mass stars, where the helium gets mixed throughout, this is the end of the road for fusion reactions. These stars gracefully retire as "white dwarfs"—smaller, brighter stars that gradually cool. They do have their surprises though; as Arthur Eddington put it, they're so dense that a ton of their plasma could fit in a matchbox.[9]

Medium stars have similar masses to our own Sun, and take a different route to the same end. As the hydrogen fusion reactions in their core sputter out and the temperature falls, they lose their stability. The star's core collapses in on itself until the particles themselves can get no closer, at which point the core begins to heat up. It eventually gets so hot in the core that helium itself begins to fuse into more massive elements, mainly carbon. The temperature required to fuse helium is twenty-five times what's required to fuse hydrogen. In the core's periphery, previously cool hydrogen gets hot enough to fuse too, producing yet more energy. The increase in temperature expands the outer layer of the star substantially—the Sun will eventually do this and be reborn as a red giant, gobbling up the Earth. The Sun has approximately 5 billion years of hydrogen fuel left before this happens, so don't worry too much. Once the Sun has depleted its red giant–helium burning phase, it too will retire to become a white dwarf, probably in 8–10 billion years.[10]

Bigger stars, with more than six times the mass of the Sun, undergo even more types of nuclear fusion. They can go from fusing helium into carbon, at 100 million degrees Kelvin, to fusing carbon into neon at 600 million degrees Kelvin, to fusing neon into oxygen at 1.2 billion, to fusing oxygen into silicon at 1.5 billion, to, finally, fusing silicon into iron at a whopping 2.7 billion degrees Kelvin.[*]

---

[*]The Kelvin scale is the Celsius scale +273.15. The difference is that Kelvin starts at absolute zero. At these large temperatures, the difference is meaningless.

Each time the star splutters and stops doing its old fusion reaction, it contracts, heats, and then begins another type of fusion. It's a system so wonderfully well adapted to producing successively heavier elements that it's hard not to think of big stars as factories pumping out the building blocks of everything in the universe. This procession of reactions continues until silicon is fused to make iron. This is the limit because it's the last reaction that produces energy from fusion, resulting in the most stable isotope, Fe-56. Did alchemists ever imagine that the starlight of a moonless sky was generated by vast stellar factories forging heavier and heavier elements? The truth is more incredible than they could have dreamed.

Stars that reach this stage are like a giant red onion, each layer responsible for a different fusion reaction. Right in the center of it all is an inert core of iron that no longer produces fusion energy. Once that core is produced, things get really interesting. All of the four fundamental forces are involved in one way or another. When the core reaches 1.44 times the mass of the Sun, the remaining particle pressure against collapse is overwhelmed by gravity, and within minutes, the core begins to collapse in on itself. It does so with ferocious speed, taking seconds to go from a radius of thousands of kilometers to just ten.

The collapsed core takes up a fraction of the space of the original core, leaving a big hole in the center of the star. The new, even denser core could become a neutron star: a star so dense that all protons and electrons are forced to combine into neutrons. With enough initial mass in the star, it could become a black hole—an object so dense that not even light can escape. Meanwhile, the rest of the star, its middle layers, have nothing supporting them and they begin to collapse inward too. Since the core can't get any denser, the incoming material bounces off it at a fraction of the speed of light, tearing back out again. A razor-thin 0.02 seconds after the bounce, the expanding shock wave slams into the outermost, and yet to collapse, layers of the star. Energy released from the core in the form of high-energy particles energizes the shock wave as it travels outward at about 10 percent of the speed of light, triggering explosive nuclear fusion reactions in its wake. Even fusion reactions that use energy are possible, and this is our main source of elements bigger than iron.

The expanding envelope of star material reaches far away into space from what was, moments earlier, a giant star. The brightness is 10 billion times the Sun's: a single exploding star can outshine entire galaxies. This is nuclear fusion on an epic scale and it fully deserves its spectacular name: supernova.*

Such events only happen in Earth's patch of space once in every half billion years, so your chances of seeing one are small. However, in 2020, some scientists grew excited observing that Betelgeuse—just seven hundred light-years away and previously the eleventh brightest star in the night sky—had entered into an unusual pattern of dimming that might precede its going supernova. In fact, that event *is* expected to happen soon—but only soon in astronomical time, which could mean at any point in the next one hundred thousand years.[11]

The death of stars is what has enabled us to exist. Mostly, humans are cleverly arranged bundles of hydrogen, carbon, oxygen, nitrogen, calcium, phosphorous, sodium, potassium, and sulphur. Apart from hydrogen, these elements are mostly forged in the last, dramatic moments of a star's life. We're all made of dead stars, and hydrogen. Without fusion reactions, complex life, which is based on a variety of atoms, wouldn't exist. Our relationship with stars and nuclear fusion goes deeper than even ancient Sun worshiping cultures could have suspected.

---

*There are different kinds of supernovae. Another involves a white dwarf sucking up mass from another nearby star. The different kinds synthesize different elements in their fusion reactions.

CHAPTER 5

HOW TO BUILD A STAR WITH MAGNETIC FIELDS

"We say that we will put the Sun into a box. The idea is pretty. The problem is, we don't know how to make the box."

—*attributed to Sébastien Balibar (CERN) and Pierre-Gilles de Gennes (Nobel laureate in physics)*[1]

As soon as nuclear fusion reactions were discovered, scientists realized that smashing together the ingredients for fusion—deuterium and tritium, the special isotopes of hydrogen that combine in the easiest to achieve reaction—was not a viable route to energy. They also knew that stars were really, really good at producing fusion energy. They had a conundrum: How could they make their machines starlike enough for fusion to happen?

The answer is to get the fusion fuel hot. Very hot. And so I've come to the hottest place in the solar system: a village in Oxfordshire.

Arriving is surreal. Although the station, Culham, is on the line between London and Oxford, the fast intercity trains don't stop here. I had to change trains and take a ponderous service that seemed to stop at every house and siding. When I do finally get off at Culham, I find I'm in a very small, plain Oxfordshire village: an unlikely location to ring in the future of global energy production. For a moment, I fear I'm in the wrong place entirely. Somewhat reassuring is the unusually large number of other passengers disembarking, more than you'd expect for an out-of-the-way English village. They all troop, improbably, toward an overgrown country path behind the station. It takes me a moment to realize that what I'm seeing is a stream of scientists and engineers heading for the Culham Science Centre. I follow them through the greenery.

When I get to it, I find that the Culham Science Centre is a myriad of public and private research institutions located on a disused Second World War–era military airfield. The Centre for Fusion Energy, where the magnetic confinement fusion experiment I'm visiting is located, occupies a large portion of the site. Some of the buildings show their age. Many have lackluster squares of glass alternating with dark green panels. In among the aggressive architecture, there are hints of the future. As I walk toward the building that houses Culham's star machine, several autonomous vehicles pass me, scanning the quiet roads of the facility with mounted radar. I see signs for firms with names like Reaction Engines (reusable space launch vehicles), GeneFirst (molecular diagnostics), and Neuro-Bio (Alzheimer's treatments).

When I get to the lobby of the Culham Centre for Fusion Energy, I find that it's *also* a contradiction in time—it resembles a careworn smoking room last redecorated in the 1970s, and yet vivid printed posters showing the "future of fusion" take up one wall.

I'm here to see the Joint European Torus (JET), the most successful fusion reactor in history. It's the apex of an approach to fusion that borrows the barely imaginable temperatures of stars— and goes *beyond* them—to get fusion working. JET was built and funded by European countries and is run by the UK Atomic Energy Authority (UKAEA). Although it's an EU project run by a UK government research organization, the staff, the support, and the mission are very much global—whatever is discovered is shared worldwide so that it can inform future fusion machines.

My first meeting is with Lorne Horton, who goes by the odd-sounding title of JET exploitation manager. He's going to show me the machine up close, or as up close as anyone can safely go while it's in operation. Lorne is a friendly, straight-talking Canadian engineer whose job it is to knit together the program of experiments that European countries want to do on JET with the day-to-day operations of the facility. He's wearing a suit, unlike many of the other scientists I see around the facility, and he looks back at me with small blue eyes under a crop of receding blond hair.

I want to find out what makes star builders like Lorne tick. Given that scientists and engineers have worked on fusion for decades and could have had more lucrative careers in the oil and gas industry, or even in scientific fields of inquiry with more immediate rewards, you've got to wonder what makes them go on day after day working on the Promethean challenge of nuclear fusion. Although Lorne doesn't seem like he'd be fazed by much, he's surprised that I'm interested in his motivations.

"I was always interested in energy as a problem," Lorne says. He grew up in what he describes as Small Town Canada, where his father worked on Canada's own brand of nuclear fission technology, CANDU reactors. For the Horton family, nuclear energy is the family business. I press him on what he means by energy as a problem. The global economy, the well-being of everyone on the planet—it all comes down to energy, he tells me. He has a crystal-clear memory of being at school when the first oil crisis blew through North America, spiking prices to unheard of levels. He thinks fusion is both the best solution and a great way to take the geopolitics out of energy.

Like so many other star builders who I talk to, Lorne can't help but be attracted by the gargantuan challenge of taming the stellar fires here on Earth. But he's also effusive about the other aspects of working on a big, international scientific program. Because JET is supported by so many countries, there's a global community working on it that he enjoys being part of. Not that there isn't healthy competition with international colleagues *not* at JET, he says.

Next, he leads me down to the engine room of star power, what he calls the diagnostic hall. It's a vast warehouse-like structure. I hear the unceasing whirr, clank, and buzz of machinery: the air pumps, cooling systems, heating systems, experimental and safety diagnostics, and the hundreds of other mechanical arteries that feed this star engine.

The large volume of the hall is subdivided into rooms with thin walls and plastic windows. They fill the space with confusing alleys. Lorne strides through them. At the center of the hall is the thick bunker that houses JET's reactor chamber, a thirty-meter-sided (approximately one-hundred-foot) cube of two-meter-thick (approximately six-foot) concrete. On the inside, there's a boron-rich scree to absorb stray fusion-produced neutrons; otherwise they'd hit regular materials and could make them radioactive. The whole reactor-containing cube is depressurized to further isolate any isotopes that they do not want escaping into the atmosphere. It can only be entered through an airlock. Any significant change to the reaction chamber means moving a set of giant concrete slabs, nine hundred tons apiece, that serve as doors and are set into the concrete cube.

As Lorne is explaining all of this to me, we pause while a countdown crackles out over the public address system. This is science at extremes, but it's so normal for star builders like Lorne that when the countdown is on, he just waits patiently. "Ten," the voice says, nine, eight, and so on, and suddenly, I'm standing just a few hundred meters away from what is—for a few seconds—the hottest place in the entire solar system.

150 Million Degrees

Inside JET, the temperature just hit 100 million degrees, substantially more than the 15 million in the Sun's core. Those high temperatures are just what's necessary for fusion to be possible on JET, Lorne says. On the face of it, there isn't much difference between a cold object and the same object but hot. They're both made from exactly the same material, and more often than not, they look the same too. So what changes when objects get hot?

Temperature is a measure of average energy, and that in turn describes how fast particles are moving around within a material. Imagine a schoolyard full of children pretending to be particles. The temperature scale has a minimum, called absolute zero. This is equivalent to the children staying still like statues, improbable though that might be. Now imagine that the children begin to run around until they bump into one another and change direction. This is what particles are doing as you add energy, and the temperature rises. The more energy, the higher the average child's speed, and the higher the temperature. High speeds and energies increase the chances of two particles—or children—colliding. When two particles (but hopefully not children) collide with enough energy, there's a chance that nuclear fusion will happen. Below a threshold collision energy, there's almost no chance. The temperatures on JET are equivalent to an average deuterium speed of more than 1 million meters (approximately 3 million feet) per second—enough to make fusion more likely, but not guaranteed.

Making a big batch of fusion fuel hot at once creates lots of chances for fusion: if one high-energy collision isn't successful, it doesn't much matter, because the next one might be, or the one after that. When enough fusion reactions are happening, the energy released can be enough to keep the reactor going. When the reactions become self-sustaining, star builders call it "ignition."

Before getting to ignition, you have to keep pumping energy in to replace any energy that's escaping. It's like having a leaky bathtub that constantly has to be topped up. Lorne describes putting this energy in as applying a blowtorch to the fusion fuel. The ratio of fusion energy coming out to energy put in per second is called "Q" in magnetic fusion, and it's the primary measure of the success of these reactors. The immediate goal for magnetic confinement star builders is getting to net power gain, a Q greater than one, because it means there's more energy coming out from fusion per second than is being pumped in. But they want to do far better than that: it's possible for no external heating at all to be required, for fusion to be completely self-sustaining. In that case, Q becomes infinite: this is ignition.

There are different types of Q: when most star builders talk about Q, they mean the ratio of energy out to energy in per second for the *plasma*. But, for practical power generation, the "wall-plug" Q is important too; this refers to the ratio of energy out to energy in per second for the whole machine. For now, JET is aiming to get to a plasma Q of 0.3 to 0.5 for five seconds, which would be a new record for total fusion energy.

After leaving the diagnostic hall, we head to JET's control room. Before long, Lorne and I are looking at live data coming in from an experiment just like the one I heard the countdown for over the public address system. There are more than a dozen scientists scurrying around the computers. Contrary to popular belief, they don't wear white coats. Most are wearing casual shirts or blouses, a few T-shirts and jeans. What's important to them is science, and understanding how the rules of the universe play out. Some of the technicians are chatting, some are doing analysis, some are checking observational equipment. Most are staring at the bank of monitors arrayed in front of us. The most compelling and, it must be said, understandable of these monitors shows a direct video feed of what is going on inside the fusion reactor. Mostly, the image is dark, but we can see a wispy purple-red fog clinging to the bottom and sides of the inside of the vessel. This is 100-million-degree hydrogen radiating away energy as light. Ideally, JET would reach 150 million degrees, because that's the temperature at which it's easiest to get to high values of Q.

Such temperatures seem entirely fantastical, but they can be measured; as an earl in Britain's House of Lords once asked, what kind of thermometer do you need to measure a temperature of 100 million degrees without it melting? To which a viscount quipped, "A rather long one!"[2] The true answer is that the calculation is made by firing a laser into the reactor chamber and measuring how much the particles of light in the beam, the photons, are changed following their collisions with electrons in the plasma. The hotter the electrons, the faster, on average, they'll move in any particular direction, and the more light hitting them gets bounced around. It's like if you dropped tennis balls from a bridge down onto two opposing lanes of motorway traffic: you'd get a much wider range of tennis ball speeds from fast moving traffic than you would from stationary traffic.[*] This interaction between light and electrons is called Thomson scattering after J. J. Thomson, Ernest Rutherford's first boss in the UK.[3] Unlike the National Ignition Facility, JET only uses lasers to diagnose what's going on during experiments. The lasers don't play a role in creating the conditions for fusion.

To bring ordinary hydrogen isotopes to the extraordinary plasma temperatures required of JET isn't easy. Conventional heating methods don't work. You can't use a fire since there shouldn't be any oxygen in the chamber, and in any case, fires just don't burn anywhere near hot enough. More elaborate heating is required.

[*]Do not try this!

The star builders' first method is to push enormous currents through the hydrogen. Ever notice your phone charger getting hot if you leave it on for a long time? A current going through a wire generates heat. The same effect works in hydrogen gas—push a current hard enough through it, and it heats up. The second method is, somewhat surprisingly, radio waves—not the kind you normally listen to but ones that are specially tuned to a frequency that makes the atoms in the hydrogen gas vibrate. As they vibrate, they crash into each other and heat up. The final method that magnetic confinement star builders use to get to 100 million degrees is to fire beams of atoms into the gas at great speeds (and so great energies). These atoms crash into the increasingly hot gases, dumping their energy as they go and heating what's already in the chamber even further.

The amount of gas that they make fiercely hot is very small—if you scooped up all of the particles and weighed them, they'd have a mass of just 0.1 milligram (approximately 35 millionths of an ounce). This makes for a gas density in the chamber that is just one-millionth that of air. The Sun is many, many times more dense than this (which means more collisions); JET gets even hotter than the Sun partly to make up for its much lower density of particles.[4]

In an energy-producing experiment, that fraction of a milligram would be made up of equal parts of the two special isotopes of hydrogen that it is easier to make fuse, deuterium and tritium. JET is an experimental reactor, built to explore the physics of nuclear fusion. It was never designed to produce electrical power for the grid. Because of that, most of the time JET uses deuterium alone as it's easier to handle and much more common than precious tritium. Physicists can still learn a lot by running with deuterium alone because most of the problems with reactors on the front line of net energy gain come down to the plasma, rather than the fusion reactions themselves. And deuterium does fuse with itself, so they can still record nuclear reactions and see some energy being released;

it's just a much less likely reaction than the one between deuterium and tritium.

Trapping Hot Hydrogen

As we've already seen, the three keys to net energy gain are temperature, density, and—finally—confinement. Once the star builders here have filled JET with hot hydrogen plasma, they need to keep it that way. And this is the great problem facing tokamaks: most machines can only run for a few seconds at a time before they lose control of their plasmas.

So how do you create a trap for something that is more than 100 million degrees? Most people's first instinct is that the hot deuterium (and sometimes tritium) is trapped by the metal walls of the reactor chamber itself. But that doesn't work, not if you want to create a mini-star. If you've ever put a hot pan in water, you'll know why. There's hissing and fizzing as the heat energy drains straight out of the pan and into the water. This fizzling out is exactly what star builders don't want to happen to their fusion plasma. If all of the high-energy particles escaped into the wall, they'd transfer their energy over too, effectively cooling the remaining hydrogen and stopping the fusion reactions dead. It's not good for the walls either; they can reach 300 degrees Celsius (572 degrees Fahrenheit) during a typical experiment, but they wouldn't survive 100 million degrees Celsius (180 million degrees Farenheit).[5]

Star builders need to pull off a much more impressive trick: to stop the plasma from touching *anything at all*. It sounds impossible: they need an invisible floating box that is impervious to heat. The star builders' solution sounds impossible too: invisible force fields. But it's not science fiction, even if it sounds like it.

The trick is magnetic fields, with which star builders weave a complex web. Everything is controlled by those four fundamental forces of physics—the strong force, the weak force, electromagnetism, and gravity. Electromagnetism is responsible for electrical charge, magnetic fields, and light. The temperatures required for fusion to work create a plasma, which consists of charged particles. This makes magnetic fields a good, if tricky, way of trapping and controlling the plasma.

The idea of using magnetic fields to confine plasma was one that star builders came up with very early on in their attempts to do fusion. When a charged particle encounters a line of the magnetic field, it gets trapped and follows it, orbiting around the line of the field in loops as it goes. The size of the loop that charged particles gyrate around is named the Larmor radius after the physicist Joseph Larmor. The stronger the magnetic field, the smaller the Larmor radius, and the more tightly bound the particles are to the magnetic field lines. Because the particles are moving anyway, in addition to gyrating, the pattern they trace out is a helix—as if they were on a helter-skelter (an amusement park ride in the UK, with a spiral slide built around a tower). Stronger magnetic fields provide better confinement of plasma partly because charged particles in the plasma are kept closer to the magnetic field lines.

Regardless of the neatness of trapping particles on magnetic field lines, early machines using this approach had many problems that JET has had to overcome, or at least *attempt* to overcome.

Perhaps the first serious magnetic confinement fusion design can be credited to Max Steenbeck. Prior to the Second World War, Steenbeck was the laboratory director of a Siemens-Schuckert plant and a member of the *Volkssturm*, the Nazi people's army designed to be the last line of defense during an invasion of Germany. The plans for his confinement device found their way to Imperial College, where they were refined by the son of J. J. Thomson, G. P. Thomson, who filed a patent for a doughnut-shaped, or toroidal, star machine in 1946. This was inspired by the "pinch" effect whereby large currents traveling through a conductor can, via the magnetic force, squish the conductor inward. It's this effect that causes lightning rods to become misshapen once struck.[6]

The simplest version of the toroidal pinch machine consists of an airtight Pyrex doughnut coated with metal that, when current flows through it, creates an electric field in the gas inside. This field provides enough energy to rip the atoms of the gas inside apart into glowing plasma and to push the charged particles around the doughnut's tube like NASCAR racers. Electric and magnetic fields are so interrelated that they cause one another: the current running around the ring creates a magnetic field that wraps around the inside of the ring at right angles to the current. This same effect is why a magnetic compass gets confused next to a live electricity cable.

But the really clever trick of the toroidal pinch is how the interplay of magnetic fields and current traps the plasma. The electromagnetic force dictates that whenever there is a current in one direction and a magnetic field in another, charged particles must move in the third and final direction, at right angles to the other two. Imagine that you are a particle going through the doughnut's ring as part of the current; then the magnetic field is wrapping around you in a circle, pointing clockwise. The direction of the force that's generated then points back at you from every angle. It's this force that pinches the plasma's particles together in the middle of the ring, squishing them up to get them closer to the density and temperature needed for fusion to have a chance. The name "toroidal pinch" arose because the hot fusion fuel is pinched into a torus shape: a doughnut within a doughnut created by an invisible force that can briefly hold the hot fuel in place without its touching anything.[7]

Toroidal pinches aren't the only early magnetic confinement device that star builders tried. Another early design, pioneered by scientists at Lawrence Livermore National Laboratory (where NIF is based), was, essentially, a sausage of magnetic fields, with the magnetic field lines going lengthways along the sausage. Particles would zip up and down. At either end, the magnetic field lines came together in dense bunches, like the tie in a wheat sheaf, the effect of which was to reflect particles back the other way. These early star traps were optimistically called magnetic mirrors, because they were supposed to reflect the particles. In reality, many particles leaked out of the ends—more like a polished window than a mirror.

There is a problem with these early pinch and mirror machines that goes beyond mere leakiness though—a problem that means they're unlikely to ever be good for net energy gain. JET has only partially solved it. Both toroidal and magnetic mirror star machines suffer from debilitating instabilities: catastrophic failures in which the magnetic field gets so rucked up that the confinement is completely lost.

Dr. Fernanda Rimini spends her days trying to understand the instabilities that plague magnetic star machines. She's a physicist and one of the session leaders of experiments on JET, responsible for translating the scientific ideas of her colleagues into settings that are feasible for the machine. She has a commanding role in the control room. While we're chatting in one corner, several people try to ask her questions on this or that technical detail. Despite the importance of her role in the smooth operation of JET, and my happening to catch her at the end of a shift, she's still full of energy. I don't think she ever switches off.

"Heavy metal music for me is the perfect soundtrack to an afternoon spent building with Lego," she says, when describing how she relaxes in her spare time. She has even re-created in Lego the building that houses JET and a miniature version of the control room. She says that what's kept her in star building is the curiosity and fun. "And you're doing work for the future of humankind," she adds. "It's an energy source, and it's important."

Fernanda describes her role as halfway between pure physics and engineering, and I can't help thinking that this is currently apt for the whole of fusion the world over.

We're talking in the busy control room, and I struggle to hear her over the background ruckus as months of planning come to fruition next to us. Fernanda explains that if they can control instabilities, they'll be well on their way to having a working star machine. Right now, instabilities are one of the reasons JET can only operate for a few seconds at a time.

While instabilities aren't unheard of elsewhere in nature—take, for example, an out-of-control garden hose or the snowball that grows into an avalanche—plasmas are especially susceptible to them, to the point where the disruptions caused by plasma instabilities frequently stop machines entirely. Instabilities arise in fusion because nature hates extreme conditions in temperature and density as well as magnetic and electric fields existing in a small patch of space or time. Given a chance, nature will even out such inequalities. Because there are such extremes of energy involved in fusion, the reestablishment of balance can happen in abrupt and surprising ways, like a stretched elastic band suddenly being released.

Professor Jerry Chittenden of Imperial College London has said that attempts to confine the fuel for fusion are subject to Murphy's Law: anything that can go wrong, will go wrong. Nuclear fusion is flighty and all too easy to stop—one way in which it's far harder than nuclear fission (and also one reason why it's safer).

Those early toroidal pinches suffered two instabilities in particular when run with higher currents. One is the kink instability, a devastating fusion-stopper that JET's clever design mitigates. It's the kink instability that causes the breakup of the hoops of plasma thrown out of the Sun's surface. In toroidal pinches, kink instabilities occur because the magnetic field is stronger on the hole side of the torus's loop than on the outer side, since the charged particles are closer together on the inside curve. But because the magnetic field is weaker on the outside of the curve, the pinching effect is weaker too. This allows some charged particles to bulge out, weakening the field further. The weaker field permits even more charged particles to bulge out, and so on, until the process runs away with itself, confinement is lost completely, and fusion stops dead. The second problem with toroidal pinches is the sausage instability. This amplifies tiny imperfections in how much the column of plasma is pinched, causing a shape like a chain of sausages to form and to ruin confinement. While JET is less prone to instabilities and the disruptions they cause than earlier magnetic star machines, they're still a big problem.[8]

JET is a tokamak, the most popular and successful type of magnetic confinement fusion machine. Even among tokamaks, JET is famous because it holds the world record for the highest Q (the ratio of power out of the plasma to power put in) ever achieved, 0.67. The machine itself is impressively big: the reaction chamber stands around six meters (approximately twenty feet) tall and three meters (approximately ten feet) wide. It's shaped like a fat doughnut, with a tube that's two meters (approximately six and a half feet) in diameter.

Tokamaks were first developed in the Soviet Union in the early 1960s, the name derived from the Russian terms *toroidalnaya kamera* and *magnitnaya katushka,* meaning "toroidal chamber" and "magnetic coil." The genius of the Russian star builders, and what allowed them to stop the kink instability from being so debilitating, was to introduce a second magnetic field to their toroidal chamber.

Figure 5.1 shows the basic design elements of a tokamak, including the reactor chamber. Inside, there's the ring-around-the-doughnut field, known as the poloidal field (one example loop is shown with a solid line and arrows), that was also present in the toroidal pinch. As in the toroidal pinch, this magnetic field can be generated by the particles' own motion. But there's a second field too (shown with a dotted line). This is the field that the Russians added. The second field, which goes around the doughnut's tube like the particles do, is called the toroidal field because it has the shape of a torus. The clever part of the Russians' design is that the two magnetic fields combine to create an overall magnetic field that twists around the tokamak helically, like a slide wraps around a helter-skelter, but on the inside of the torus.

Particles are confined to gyrate around the helter-skelter magnetic field lines. Because they move from the inner to the outer part of the doughnut's tube as they go around it, particles spend as much time on the inner side of the tube as they do on the outside.

A schematic of a tokamak

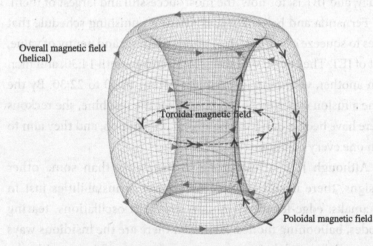

Overall magnetic field (helical)

Toroidal magnetic field

Poloidal magnetic field

Figure 5.1: The basic design elements of the tokamak, the leading magnetic confinement fusion star trap. The reactor is shaped like a doughnut. Two different magnetic fields combine to make a helical field (shown with filled-in arrows) that wraps around the inside of the tube of the doughnut-shaped reactor chamber. Charged particles (electrons and hydrogen nuclei) get trapped on the magnetic field lines, gyrating as they travel along them. Magnetic confinement keeps these particles from hitting the chamber walls—sometimes.[9]

That stops the bulges endemic to toroidal pinches from developing in the first place, which prevents them from growing, which keeps the plasma flow uninterrupted.

Russian tokamak scientists suppressed the kinky tendencies of the toroidal pinches that had gone before. Straightaway, their tokamak design provided tenfold better confinement than alternative magnetic schemes of the time. The strategy was so successful that, initially, no one outside of the USSR would believe how good the results were. To settle the question, a team of scientists from Culham was sent to Russia at the height of the Cold War to use Thomson scattering as a more precise check on the temperatures. The Russians were right; tokamaks were confining hot fusion fuel like no other machine before them. Tokamaks soon became the leading magnetic star traps the world over, and they remain so today, and JET is, for now, the most successful and largest of them.

Fernanda and her colleagues follow a punishing schedule that tries to squeeze every last bit of understanding, and penny of value, out of JET. They begin one shift at 6:30 and go until 14:30, and then run another, with more experiments, from 14:30 to 22:30. By the time a fusion experiment is prepared for the machine, she reckons there have been around six months of preparation, and they aim to run one every half hour.

Although JET is less prone to instability than some other designs, there is still a whole menagerie of instabilities just in tokamaks: edge-localized modes, sawtooth oscillations, tearing modes, ballooning modes. And then there are the insidious ways that particles, and their energy, try to escape confinement, like the bizarrely named "banana regime transport" (after the shape the particles trace out). These are driven by energetic collisions that send a particle spinning out of the web of magnetic fields. Each different way for particles to escape presents a different headache for star builders, and there are pages and pages of mathematics that try to understand how they happen, and what scientists can do to mitigate them.

Today, Fernanda is coordinating experiments that look at how injecting pellets of frozen fuel at just the right moment can stop the instabilities from killing the confinement. They do it by cooling down a portion of the fuel.

When a tokamak runs into a bad instability, it can cause a catastrophic loss of energy called a disruption. In fast disruptions, the energy in the fuel is violently dumped into the vessel in a few milliseconds.

"You blink and your fuel is no longer there," Fernanda says. The effect of this is to make the whole machine strain at its moorings, with forces equivalent to hundreds of tons and jumps of a centimeter or more. For this reason, JET rests on giant springs.

Just as seriously, a sudden energy release can melt the walls—even though they can withstand a few thousand degrees. Unleashed electromagnetic forces are strong enough to twist the reactor structure, misshaping it and ruining its confining properties. Disruptions are bad news.

Fernanda says that predicting disruptions is a big physics challenge facing tokamaks. One large disruption could terminally damage the reactor. Lorne Horton told me that industrial partners who might actually build a fusion reactor are adamant that the technology must be predictable to have a chance of being commercially viable. A reactor that stops without warning, potentially breaking itself in the process, is just not going to allow star builders to achieve their ambitions.

It's no coincidence that Culham's star-maker-in-chief, Professor Ian Chapman, made his (scientific) name by trying to pin down the causes of instabilities in tokamaks. He's published more than 110 scientific papers and has been invited to speak to other physicists at conferences all over the world. His speciality is the sawtooth instability; it causes temperatures and densities to slowly climb, only to suddenly crash back down again, firing off hot plasma into the reactor walls as it does so.

Disruptions can even threaten the quality of the energy confinement of the reactor in future experiments. When particles from the plasma strike the wall and melt it, even on a microscopic scale, heavy wall materials are released into the fusion fuel. This heavy metal is definitely not Fernanda's favorite, as it acts like a pollutant and has a terrible effect on the ability of the reactor to confine energy.

JET's internal walls are mainly made of beryllium and tungsten, atoms that have four and seventy-four electrons respectively, compared to a hydrogen atom's one electron. Because deuterium and tritium just have one electron each, and because it's so hot in tokamaks, they get completely dissociated from their nuclei when they turn into plasma. It takes much more energy to remove all of the electrons from beryllium and tungsten, and that is precisely why they cause problems when they pollute the plasma. As beryllium and tungsten become partially stripped of their electrons, they suck energy away from the hydrogen fuel.

Worse, the electrons that remain attached to the beryllium and tungsten nuclei can temporarily go into an excited, higher energy state. When they relax again, the energy of their excited state has to go somewhere, and it tends to come out in the form of light. This light is an energy escape route straight out of the hot plasma.

The pattern of colors that the transitions of electrons create as they jump into different states are like fingerprints; they say what particles are present, how many electrons they have, and what the temperature and density around them is. This is why a hot flame burns blue while a colder flame burns red, or even yellow. The pattern of colors tells a story about what's going on in the hot fuel—which we know is too hot for any thermometer to work—and star builders often use this technique of looking at the light emitted, called spectroscopy, to infer conditions in their star traps. The light radiation from atoms whose electrons are partially stripped away can be useful. But when metals are radiating away energy from the center of a tokamak, it's really bad news.

Even tiny amounts of wall material getting into the fuel ruin the reactor's performance. At less than 0.02 percent tungsten, the radiation emitted in the form of light carries away a whopping 50 percent of the energy from the helium nuclei created in fusion reactions. Elements with more electrons light up like Christmas trees.[10]

For this reason, most of the walls of JET are made of beryllium, with only four electrons. Only the points where the very worst, and hottest, fusion fuel lashes, whiplike, into the walls are made out of the more robust tungsten. Tungsten has the highest melting point of any metal, at 3,422 degrees Celsius (a little less than 6,200 degrees Fahrenheit), while beryllium melts at a comparatively chilly 1,287 degrees Celsius (approximately 2,350 degrees Fahreneheit). A Q greater than one tokamak would see the part of the wall specifically designed to take the biggest pounding, the divertor, exposed to a heat load equivalent to that of a spacecraft reentering Earth's atmosphere. Truly, star building requires mastery of extremes that make bringing a shuttle down from orbit seem a cinch. The design of tokamaks has cleverly incorporated the divertor to be the one place that takes an absolute beating. It also serves, as the name suggests, as a place where impurities that flake off the main beryllium walls are isolated and removed, leaving the fuel in the doughnut's tube pure.

Purity is so important that the cleanliness of the inside of the tokamak chamber would have even Marie Kondo impressed. There must be no air, no water, no grease, no oil—nothing that could contaminate the fusion fuel and unwittingly cause energy to leak away as light radiation. The vacuuming of a tokamak is so effective in removing particles that it's equivalent to removing all but a single star from the Milky Way. The scrub job is made harder because atoms like oxygen don't just float around in molecules, they can get trapped in the walls too. The only way to deep clean the walls is to bake the chamber at a few hundred degrees for long periods of time. Because of this, Lorne Horton tells me that to do anything in the chamber with humans, which inevitably involves a cleanup, takes a minimum of three months, without counting the time taken to actually do the work.

The need for such care has also seen the Culham Centre for Fusion Energy become a center of excellence in robotics. Remotely controlled robot arms can get in and out without disturbing the chamber's vacuum. While Lorne and I were wandering around Culham's main building, I saw trainee engineers trying out the remote-handling arms for the first time via a controller that looked like a periscope with joysticks. Later, in a mock-up of the reaction chamber housed in a corner of the diagnostic hall, I saw the engineers' every move translated into the surprisingly graceful movements of a metal arm that weighed tons. The sense of touch created by the robotic arm is so good, Lorne told me, that experienced operators can feel if a bolt is not aligned with its thread.

The robotics technology is being spun out separately to all kinds of industries where remote handling might be useful, including nuclear fission. In one warehouse on site, I saw a three-storey wire mesh cage with robotic arms busily completing tasks inside; another had drones angrily buzzing back and forth. The scientists here really are quietly creating the future.

The World's Most Successful Fusion Experiment

The world is paying attention. JET is, after all, a collaboration between many of the nations of Europe under the name of Euratom. Although the idea was born early in the 1970s, it took years of wrangling before a location was agreed upon. Governments jostled to host the facility that would be a shining example of technological progress and would undoubtedly bring with it good jobs and lucrative contracts. Six different sites were fought over. "Mad" Rebut, Paul Henri-Rebut, the driving force behind JET's design and eventually its director, was so incensed by the delay that he suggested putting JET on the *Queen Elizabeth 2* ocean liner so that it could dock in a different European port each month. Finally, in 1984, JET was completed and opened by Queen Elizabeth II (note, not *on* Queen Elizabeth II) and then French President François Mitterrand.

Ian Chapman tells me that JET cost about £2 billion (a little more than $2.5 billion) in today's money and took four years to build. You might wonder what Culham's star builders have been doing with it since it opened. Surely, a machine that is more than thirty-five years old can hardly be at the cutting edge of technology, much less usher in an energy revolution.

JET set the world record for plasma Q with deuterium-tritium fuel in 1997, at 0.67—that is, for a brief moment, 67 percent of the energy put in by Lorne Horton's blowtorches was released by fusion reactions within the reactor chamber. It produced sixteen megawatts of fusion power; very roughly, that would be enough to power thirty thousand homes if it was available continuously.

This seems so tantalizingly close to a plasma Q of 100 percent that you're probably wondering why, in the years since, they haven't managed to get any further than they did in that single landmark experiment. The truth is that this was on the edge of what JET could do in the 1990s and 2000s, and more important, the energy output was only maintained for less than one second because of instabilities that ruined the confinement of the plasma. The 100 percent barrier is very important psychologically, including for those who are funding fusion, but reaching it doesn't automatically solve fusion, especially with such short running times. There's also the subtlety that only one-fifth of fusion energy comes out as a charged particle. The neutrons that carry four-fifths of the energy aren't charged and so fly right out of the magnetic trap. The reactor *chamber* will lose energy without a Q of at least five (though these neutrons are ultimately what are used to generate electricity, so their energy is critical in other ways).[11]

Perhaps JET's greatest success wasn't getting to a plasma Q of 0.67 for a brief moment but getting to a Q of 0.18 for five seconds. For commercial fusion energy, tokamak power needs to be sustained for hours, if not indefinitely. JET has been upgraded, repaired, and modified several times over the years, each time missing action for months and months. Having been refitted, it is about to gear up for a new campaign of experiments with deuterium and tritium. It remains the only tokamak in the world able to run deuterium-tritium experiments. Star builders like Lorne Horton, Fernanda Rimini, and Ian Chapman have a good chance of smashing JET's old record for fusion energy, and a successor to JET that aims to exceed a Q of 1 is already under construction. But it will only happen if the star builders can control the writhing plasma in their fusion furnace.[12]

You may wonder why star builders are so optimistic that they *will* get net energy gain. It's not just because of JET's exciting experimental successes; the first fusion devices were in the 1940s, and the tokamak didn't come along until the 1960s—both long before JET was in operation.

Star builders have believed in fusion for many years not only because of successes in experiments, but because of a great success in theoretical physics. Just as theory has illuminated nuclear physics, and clarified to star builders what reactions are possible in what conditions, so theory has given guidance on what conditions star machines need to create in order to achieve net energy gain.

Way back in 1957, a plasma physicist named John Lawson

proved a theory that shows that there is no physics-based barrier to igniting fusion reactions in a star machine. "Igniting" here means getting to conditions where fusion sustains itself through its own energy release, like a star.[13]

His argument was beautiful in its simplicity: he said that, at a minimum, more energy must be coming into the plasma than leaving it. He then added up all of the ways that energy could *get into* the fusion plasma, including heating from external sources, like Lorne Horton's blowtorches, plus the energy of fusion-produced helium nuclei (he assumed that the fusion-produced neutrons would all escape because they can't be confined by magnetic fields). Next, he added up all of the ways energy could be *lost* from the plasma, including through bremsstrahlung radiation (X-rays generated by the movement of charged particles in the plasma) and the loss of any plasma that managed to escape confinement. Then Lawson looked at the energy in and energy out every second, on both sides: gains and losses. He realized that all of the gains and losses were somehow dependent on just three properties of the plasma: temperature, density, and the time period over which particles in the plasma remain confined.

Lawson knew that deuterium and tritium fusion was the easiest possible reaction for star builders to aim for, so he plugged in the numbers for that. Then came the moment of truth. Lawson might have found that there was no combination of temperature, density, and confinement time that could make fusion work. That would have suggested that star builders should just give up trying; their experiments might have improved, but they'd never reach ignition or net energy gain. But Lawson didn't find that; what he found was that there *were* combinations of temperature, density, and confinement time that would make fusion work. Lawson's equation says that star power on Earth *will* work. A tokamak that could reach temperatures of more than 100 million degrees Celsius (180 million degrees Fahrenheit), densities of more than 10,000 billion particles in each cubic centimeter, and energy in the plasma that is confined for more than 100 seconds would hit his requirements and ignite its fusion plasma.

What this theory means is that it's possible to re-create star power on Earth: hard, yes—the conditions required are extreme, more extreme than anything else on the planet—but possible. The implications of this are as enormous today as they were then. It's this insight that has kept generations of star builders optimistic about delivering on the promise of nuclear fusion: they know it's scientifically *possible*; they just need to build a machine good enough to make it happen. And JET has come very, very close.

Lawson's equation also shows why stars are fusion reactors that just keep on going; gravity can achieve high temperature, high density, and sufficient plasma energy confinement in perpetuity. More important for fusion on Earth, using Lawson's equation, star builders know exactly what they have to aim for in their star machines. Tokamaks use high temperatures, with plasma confinement times measured in seconds, perhaps eventually in hours, thanks to using magnetic fields. But Lawson's equation also tells us that magnetic fields are not the only way to trap a star . . .

CHAPTER 6

HOW TO BUILD A STAR WITH INERTIA

"... like trying to confine jelly with rubber bands."
—*Edward Teller, describing trapping*
plasma with magnetic fields[1]

You don't have to be in California for long to see why the two scientists most responsible for establishing Lawrence Livermore National Laboratory—Ernest O. Lawrence and Edward Teller (the controversial physicist best known as the father of the hydrogen bomb)—were happy to set up shop here. On a February morning that really ought to be cold, there's so much sunshine that I remark to the person taking me between buildings on Livermore's sprawling site that I feel like I should be outside enjoying it instead of poking around nuclear facilities. She tells me that she felt like that too, for the first year after she moved from South Dakota. But then she realized that it was sunny and fine almost every day here on the east side of the San Francisco Bay, and that she no longer needed to rush outside to catch it. It would be there waiting for her whenever.

And, as I arrived at the laboratory, I couldn't help but notice a mile-long array of solar panels soaking up that bright Californian sunshine. It's here that scientists are using laser light to create their own sunshine, via the National Ignition Facility (NIF), the world's most successful inertial confinement fusion experiment. In the hills above Lawrence Livermore, I see more evidence of their pursuit of clean energy, in the form of a sparse forest of wind turbines ponderously revolving under a laid-back breeze. If the juxtaposition of so bucolic a scene with a government laboratory bristling with security, nuclear secrets, and scientific experiments should have seemed awkward, it didn't. On site, there are reed-ringed ponds and bicycle lanes. As with most of America, cars and roads are ubiquitous too— I'm on foot, but you're almost obliged to drive around, if not due to the sheer size of the facility then owing to cultural norms. On a previous visit I was asked to drive to a building where I was scheduled to have a meeting. Following my minder, I did as I was told, and we drove down a couple of small roads to get to another parking lot. When I emerged from the car I could see, less than one hundred meters away (a little more than three hundred feet), the spot that I'd been parked in moments before.

I'm here on another visit to Livermore to find out how scientists at the NIF are using inertia to build a star. Some of the scientists working here argue that although their machine has not yet matched the record energy gain of JET, their inertial confinement fusion machine is the closest in the world to achieving net energy gain. And they *are* catching up fast.

Inertial confinement fusion is very different from magnetic confinement fusion; for a start, magnetic confinement tends to run continuously, like those takeout pizza ovens with conveyer belts, while inertial fusion is a batch process, like a bread oven. But the biggest difference is that inertial confinement fusion doesn't use magnetic fields to confine the plasma. Instead, it uses inertia. You'd be forgiven for not having a clue what that means. CEO of inertial confinement firm First Light Fusion Dr. Nick Hawker later put it to me this way: "There's nothing holding the plasma together—no big magnets, or external forces."

But surely *something* needs to hold the plasma together while fusion is happening.

Lawson's equation for net energy gain tells the star builders that nuclear fusion schemes need a combination of high temperatures, high densities, and good energy confinement to reach net energy gain. But it also tells them that they can mix between these three, like DJs mixing tracks but keeping the overall sound volume the same. Magnetic confinement star builders use plasma far less dense than in the Sun, even less dense than the air we breathe, but they do it by creating the hottest temperatures in the solar system, *hotter* than the Sun, and by using long energy confinement times.

Inertial confinement fusion uses a different mix of temperature, density, and confinement. Comparing NIF, the world's leading inertial confinement fusion machine, with JET, the world's leading magnetic confinement machine, shows the compromises. NIF's plasma reaches temperatures that are a shade less hot than on the JET tokamak. But the densities and pressures are far higher. Assuming you're reading this on Earth, you're experiencing a pressure of just one atmosphere, but the center of the Sun has a pressure beyond 100 billion atmospheres—and that's what NIF has too, its director, Dr. Mark Herrmann, told me. "So the centers of our implosions are like star matter."

But the biggest difference between inertial and magnetic confinement is how long the fusion plasmas are maintained: as opposed to magnetic fields holding plasma in place in perpetuity, the inertial approach traps it for just tens of billionths of a second, the length of time that the plasma's own inertia holds it together. At NIF, that inertia comes from the momentum of the inward explosion, or implosion, of the fuel on itself.

The implosion of the fusion fuel provides confinement, nothing else, and that only for the briefest of moments. But the briefest of moments is a long time in nuclear physics. When you pop a water balloon, the water retains the shape of the balloon for a few fractions of a second before it turns into a big wet mess. It's the same with plasma. With nothing to hold the shape of a ball of plasma, inertia holds it there for the short time it takes a sound wave to cross the sphere of plasma. But fusion happens on timescales a thousand times shorter. Nick Hawker describes inertial confinement fusion plasmas as "transient phenomena, these very high-pressure, high-temperature, high-density states of matter, and they're held together for a very short amount of time, simply by their own inertia. Once assembled, it just takes a certain amount of time for it to disassemble."[2] Each implosion is a race against time.

In the prologue, we saw this race play out in a shot fired by NIF's 192-beam infrared laser. Those beams are turned into ultraviolet laser light by crystals as they enter the reactor chamber and are focused on two small holes on either end of a small gold box, the hohlraum (German for "hollow room"). The beams hit spots on the inside walls of the hohlraum, generating high-energy X-rays. The X-rays bathe a small capsule of fusion fuel, expanding its outer layer and triggering a rocket effect whereby the remaining fuel capsule contracts rapidly. The inward explosion, or implosion, squeezes the fusion fuel into a ball with a radius the width of a human hair. With just the right combination of timing, laser pulse, hohlraum, and capsule, the fuel becomes a plasma with the right temperature and density for fusion. The fusion reactions radiate out from the center of the plasma like a wave, and engulf the rest of the fuel.

It's a series of extreme physics events strung together to make a spectacular finale. With a massive input of initial energy, fusion fuel can be collapsed in on itself for a moment of shining glory. At NIF, this all begins with the laser beam—which is fitting, because lasers are a big part of how this approach to fusion got started.

The Light Fantastic

Inertial confinement fusion began shortly after Edward Teller and his colleagues demonstrated the first hydrogen bomb in 1952. Teller had been contacted by politicians in Washington who had a new challenge for the brilliant but controversial physicist. "No sooner was it done," Teller has said, "than every politician and every bureaucrat descended upon us saying, 'Now you must solve the problem of controlled fusion.'"[3]

Teller was skeptical of the magnetic confinement fusion schemes on offer in the pre-tokamak era. In 1954, he delivered a damning lecture in which he said that all magnetic schemes that had been tried were highly susceptible to plasma instabilities, the runaway processes that ruin confinement. He had an alternative. The plasma in magnetic fusion conditions is like the solar corona, the outer, wispy part of the Sun that becomes visible during solar eclipses. Physicists still don't properly understand it, and even then Teller knew that working with similar plasmas would present tremendous difficulties because there's so much going on: heat transfer between particles, light carrying energy away, and magnetic fields bringing all kinds of complex instabilities into the mix.[4]

However, Teller thought, the inside of the Sun, where the density is much higher, is easier to model. The plasma is very pure: all electrons are stripped from helium and hydrogen, and the densities are high enough to make the plasma opaque to light. This plasma is much more similar to what is created in a hydrogen bomb, something that Teller understood better than anyone. So Teller had an idea.

In 1957, a young scientist named John Nuckolls got the job of turning Edward Teller's idea into a reality. Nuckolls's instructions were crazy. He was asked whether it would be feasible to hollow out a mountain and drop hydrogen bombs in it to generate electricity. What wasn't so crazy was that hydrogen bombs could already deliver huge net energy gains. But (and this will shock no one) exploding a series of hydrogen bombs has downsides. The mountain would eventually crumble into a radioactive mess, and the use of large numbers of nuclear weapons was an enormous nuclear proliferation risk.

Nuckolls instead had the idea of taking the hydrogen bombs out of the design altogether. This got rid of the fission reactions, eliminating much of the radioactivity, as well as the proliferation risk. Instead of a few big explosions, he'd do many miniaturized fusion implosions using just ten milligrams (approximately four ten thousandths of an ounce) of fuel each time. The smaller yields of energy would be easier to deal with too: Nuckolls would replace an unmanageable "BANG!" with a controllable "pop"; instead of a single large release of energy, he'd have the setup work like a gasoline-powered engine, with a stream of pops as a succession of fuel capsules went off.

It was a good idea, but it would only work if there was a way to compress and heat the tiny ball of fuel enough to initiate fusion. But no one had ever driven a fusion implosion, no matter how big or small, *without* using fission explosives.

"The driver may have kilometer dimensions but must concentrate energy in space and time to energize a tiny sub-centimeter-scale radiation implosion," Nuckolls wrote. "For power production, the driver focusing mechanism must be separated a safe distance from the fusion explosion. The driver must ignite billions of microexplosions in a thirty-year power plant lifetime."

Nuckolls considered various schemes to trigger fusion in his capsules, including firing jets of plasma at them, smashing them with a very fast pellet gun, strafing them with charged particles, and hitting them with exploding foils. Nothing, however, was quite right. Livermore's weapons designers weren't impressed, calling

Nuckolls's prolific series of classified memos on small fusion capsules "Nuckolls' Nickel Novels."[5]

Then, in 1960, a scientist named Theodore Maiman announced the creation of the first laser. Lasers are used in everything from the fiber-optic communications that enable the Internet, to eye surgery, barcode scanners, cleaning, playing DVDs and CDs, and cutting metals in industry. Although they're ubiquitous, there's extraordinary physics going on when a laser "lases." The light that we see most of the time, from the Sun or a reading lamp, is made up of many different electromagnetic waves. The distance between the crests of light waves is what determines their color: pure sunlight has many colors, while sunlight reflected from grass is dominated by green light. But even if light appears in the wavelength of one color only, like the light from a red lamp, the crests and the troughs of that light aren't all in sync. Laser light is special because the waves are the same shape, and so the same color, but they're also in sync with each other. The difference between colored light and laser light is the difference between a crowd walking haphazardly with the same stride length, and an army marching forth in time together.[6]

Because the light is in sync, a little laser light goes a long way. An ultramodern and efficient LED lamp might draw 5 watts of power and wouldn't do you much harm if you glanced at it. Contrast this with a laser, which at just 0.01 watts can do serious damage to your retina if you look at it directly—and it's *never* safe to look into a laser beam. The magic of lasers is that, with a lens, they can cram a lot of in-sync light energy into a tiny spot. They can focus energy in time *and* space. When Maiman first demonstrated the laser, it was the archetypal technological solution looking for a problem. John Nuckolls had just the problem.

Nuckolls realized that lasers could drive enough energy into his fusion targets. Barely a year after their discovery, Nuckolls excitedly wrote a memo to the director of Livermore, explaining his idea of a laser driven "Thermonuclear Engine," "the fusion analogue of the cyclic internal combustion engine."[7]

More than five decades later, Nuckolls's idea of using lasers to create mini-stars led to the construction of the National Ignition Facility. It's not the only way that star builders are doing inertial confinement fusion, but it's the way that has come closest to achieving net energy gain.

To find out more about how Nuckolls's idea has been translated into reality, I'm heading to the Optics Facility, a building dedicated to keeping NIF's thirty thousand or so lenses, mirrors, crystals, and laser glass in good working order. These elements are key to delivering energy, via the laser, to the fusion fuel.

As my guide, Operations Manager Dr. Bruno Van Wonterghem, leads the way, he points out where Edward Teller himself used to drive around the Lawrence Livermore site in his golf cart. I can imagine Teller, zipping along the quiet roads, saying hello to colleagues with the Hungarian accent that he never completely lost, despite living in the US for most of his life.

When we arrive in the facility, we meet Dr. Tayyab Suratwala in a corridor full of scientific posters showing detailed schematics of experiments splashed with the bright laser colors of green, ruby, and violet. Tayyab is the program director for optics, the science of light and glass and how they interact. What that means here at NIF is that he is responsible for ensuring that the parts of this gigantic laser that manipulate light energy keep working. As well as directing NIF's optics operations, Tayyab has found time to write more than one hundred scientific papers, file six patents, and write a book, all on optics.

NIF, Tayyab tells me with warranted pride, is the only laser system that continuously operates above the damage threshold for the nine thousand or so more delicate optics directly in the laser's path.

What this means is that NIF's laser is so energetic that it damages its own optics every time it fires.

The damage manifests as frequent chips, fractures, and cracks in the most technologically advanced optics in the world. Because they're so specialized, it can take an entire year for replacements to be manufactured and imported—precious experimental time that NIF's scientists would lose if they waited around. So, instead, Tayyab and his colleagues have adopted a different approach: invent better optics (easy, right?) and, rather than avoid damage altogether, incorporate damage as a regular part of running the laser.

Tayyab says that a shot by NIF with the quality of optics available back in 1997, when NIF was designed, would have created fifty thousand damage sites per optic, seeded by defects as small as ten-billionths of a meter within the glass. By inventing more perfectly polished optics, NIF's staff has dropped that to fifty damaged sites per optic. And when they do see damage, they have amazing ways of repairing it, which Suratwala shows me. I watch, astounded, as a machine learning–driven robot takes thick slabs of glass, scans them for fractures, and then uses a small but powerful laser to smooth any sharp pits into gentle valleys that mean the optic can still function well.

Even after the beams have entered the chamber, the risks to the optics aren't over. "The second most energetic laser in the world is the scattered light from NIF," Tayyab says. Any laser light that isn't perfectly carried along the beam lines causes havoc elsewhere. And even when the light hits what it is supposed to, it doesn't always *do* what it is supposed to do.

The behavior of light and plasma is mostly determined by the electromagnetic force, and the two can interact, often in unintended and unhelpful ways. To name but a few that manifest with plasmas and laser beams, there's cross-beam energy transfer (the plasma takes energy from one laser beam and puts it in another), self-focusing (the plasma acts like a lens), filamentation (the plasma extrudes smooth beams into strands like spaghetti), stimulated Brillouin scattering (the plasma acts like a mirror), fast electron heating, and a whole range of laser-driven instabilities.

Even in Nuckolls's day, the designers at Livermore knew that shining lasers directly onto a fusion capsule would risk all of these interactions complicating an implosion, and send you straight back into the realm of difficult plasma physics. When told about Nuckolls's original scheme, Teller is said to have asked "Wait a minute! Are you telling me that laser fusion involves *real* plasma physics?"—to which the presenter responded, "Yes, sir, it does." Reportedly, Teller replied, with some despondency, "Well . . . it will never work."[8]

Today, "real" plasma physics still presents a big challenge for laser-driven inertial fusion energy. But some of the problems are side-stepped when an intermediary is used to absorb the laser energy. That's why Livermore's scientists developed the hohlraum, the gold box into which the laser shines, entering through the two holes on each end. The bath of X-rays produced by the hohlraum when the laser hits its surface is smoother than the laser beams themselves, is less likely to cause plasma instabilities, and reduces the risk of preheating the capsule before it's compressed to high densities. To distinguish it from directly shining laser light on a target, this approach is called indirect drive and it's what's used by NIF.

While star builders haven't completely given up on direct drive laser fusion (France's smaller equivalent of NIF is still pursuing the direct drive approach), the indirect drive approach has been much more successful in getting fusion to work during the punishingly small window of confinement.[9]

Targeting Net Energy Gain

Even as the X-rays get to the capsule, there's a lot more going on than you might think to make the tens of billionths of a second a NIF implosion lasts a success.

So the next stop on my tour of NIF is the target fabrication laboratory, where I'm hoping to find out what role the targets play in this type of fusion. My guides are Becky Butlin and Dr. Michael Stadermann. Becky tells me that she caught the science bug in sixth grade when her teacher began assigning weekly mathematics puzzles, but she soon discovered that she enjoyed using that knowledge to make physical objects, and as a teen, she even set up a workbench in her garage. That desire to do science through building and creating objects has led to a career in fabricating targets for NIF. Becky and Michael talk to me about target fabrication as if it were *the* most important topic in the world. Their enthusiasm is infectious, and they have a point. As Michael tells me, the target (plus the laser pulse, he says in an aside) *is* the experiment. He's right; although NIF is in a huge building, the most important parameters from experiment to experiment are the laser pulse and the target. My guides certainly have me hooked: target fabrication *does* seem to be key to the whole contraption ever hitting net energy gain, so tell me more! "Well," Becky says, "would you like to go into the clean room?"

You bet I would. We enter a changing area. There are so many things to put on: a sterile full body suit, sterile slip-over shoe guards, sterile gloves, hood, and face mask with nose pinch. Becky and Michael are in their clean armor in seconds; I am still trying to pull on a latex glove using another gloved hand when they come over to assist me. The clean-room gear isn't too comfortable, and by the time they've finished, there's barely an inch of flesh showing on any of us. It doesn't help that I'm slightly nervous. Truth be told, I'm an applied mathematician at heart, and I've always tried to dodge laboratory work in favor of working with pen and paper or a computer. Given the submillimeter precision going into the laboratory's products, I'm painfully aware that I am an overenthusiastic arm gesture away from causing hundreds of thousands of dollars' worth of damage. Laboratory work is fiddly and hard, requiring saintlike patience. The satisfaction of having created something tangible that can help probe the most extreme conditions in the solar system is the reward. Following a Saturday spent building equipment, Rutherford once exclaimed to a colleague, "I *am* sorry for the poor fellows that haven't got labs to work in!"

We enter the target fabrication laboratory via a sticky patch of floor that captures any final pieces of the outside world. Inside, there are a few rows of scientists using machinery to minutely manipulate the tiny parts that go into a NIF target. There's a constant gentle hum of filtered air-conditioning. Because any stray particles that adhere to the targets can interfere with the demanding uniformity of the sphere of fuel, dust particles greater than five-millionths of a meter in diameter must be extracted from the air. It's a strange sort of office, and certainly the cleanest I've ever stepped into.

Michael tells me that a typical target costs $100,000 or more. The outer box, the hohlraum, is made of gold, while the outer layer of the sphere is made of diamond, two of the most expensive materials on Earth. However, these account for just a few dollars of materials cost because there's so little of either in each target. The hohlraum is only one centimeter long, with walls thirty-millionths of a meter thick. Working with such small sizes has its supply challenges. The lab's staff once needed to buy twenty grams (just under one ounce) of specialist detergent to clean a component; the firm only sold it in drums of two hundred liters (approximately fifty-three gallons).

The biggest cost of a target is human labor, because it takes hundreds of painstaking hours to build each target. In mass production, the costs would tumble, but for now, the targets are bespoke, tuned for specific experiments. Since NIF fires only four hundred shots a year, every single one counts and is cleverly designed to answer as many questions as possible. "Without targets, there is no science at NIF," Michael says.

He says that because even minute changes in the targets—and especially the capsules that sit within the hohlraum—can completely change the experiment. I'm standing in front of some of the two-millimeter-wide spheres. You can't see this with the naked eye, but they're like Russian dolls, each of the three layers precision-engineered to serve a different purpose in igniting fusion reactions.

It seems unnecessarily complicated, and you might well wonder why it wouldn't be easier just to have a ball of deuterium and tritium. Star builders at Livermore worked out early on that such a target design would be terribly inefficient. To make fusion happen, the X-rays would have to raise the whole thing to the right temperature and density, which means putting a lot of energy in up front. It would be like making a fire by raising the temperature to the point where every log burst into flames. Using a simple blob of deuterium and tritium like this has a natural energy release limit of twenty times the energy that you put in and so is just not good enough for commercial fusion, which needs gains of thirty to one hundred.

Instead, the design of the capsules lights a match of fusion reactions right in the center, and these spread outward. It takes a lot less input energy to do this, but it does require the capsule's layers to be cleverly engineered. The outer layer is a thin sheen of diamond just 30-millionths of a meter thick. It acts as rocket fuel for the collapse of the capsule, helping to raise its density. Next, there's a 160-millionths-of-a-meter-thick solid ice layer of deuterium and tritium. This layer provides the big, heavy logs of the fusion fire and accounts for most of the mass of the capsule. Sitting at the center, and making up most of the volume of the capsule, is deuterium and tritium gas. It takes really special conditions for this core of gas to coexist with the solid deuterium and tritium layer next to it; most states of matter exist as *either* solid, liquid, gas, or plasma at a given pressure and temperature. To sustain the unusual multi-state structure, the star builders need to maintain a temperature of just nineteen degrees above absolute zero (or -254 degrees Celsius, -425 degrees Fahrenheit) until the X-rays arrive. This final layer of gas is the kindling for fusion, the place where the initial spark begins.[10]

Because of the way reactions start in a hot plasma that's formed in the core of the collapsed capsule, this potentially high-gain strategy is called hotspot ignition, and it's by far the most successful form of laser fusion.

Hotspot ignition is a great idea in principle and a thunderously difficult one to execute in practice. Somehow a match must be held to the fuel right at the center of a dense ball of plasma without disturbing its perfectly spherical geometry.

The star builders' great innovation is to use the shape of the laser pulse to ignite the hot spot. NIF's scientists introduce a series of peaks into the energy delivered by the laser over time to, indirectly, launch a series of shock waves into the capsule. With careful timing, these shock waves both dramatically increase the density of the fuel and create the hot temperatures needed in its dead center.

Shock waves aren't that common in everyday life, but once you start looking, they're everywhere. They're one of my favorite phenomena in physics. One way to think about them is as a sudden leap, from low density to high, or low temperature to high. It's no coincidence that we use the same word, "shock," to describe a sudden change of state in emotion too. In physics, shock waves appear whenever something moves through a medium faster than the waves of the medium themselves can move. Fast-moving ships can create bow shocks ahead of them in water. The space shuttle reentering the Earth's atmosphere creates one around its nose that's hot enough to turn air into plasma. Supersonic jets create them as they travel faster in air than sound waves do, and you can actually hear them as a loud BOOM-like thunder. Thunder is itself the shock wave created by lightning turning air into plasma. Explosions can create shock waves too.

A single shock wave would only push up the density and temperature of the fusion fuel so far; certainly not enough to reach the thirty-thousandfold density increase that is needed for net energy gain. So, to push the fuel all the way, three (or more) shocks are fired off in succession using laser pulse shaping. These shocks have increasingly high energies, so, cleverly, the later ones travel faster. The intended result is that their effects coalesce only once they are within the capsule's fuel. This mammoth combined shock compresses the cold, icy fuel to ten times the density of the Sun's core and creates more pressure than an aircraft carrier balanced on a penny. The capsule is shrunk to just a thirtieth of its original radius, and if all goes to plan, the fuel is engulfed by fusion reactions.[11]

Seeing Stars

To make this house of cards stand up requires everything to work perfectly together: the laser, the hohlraum, the capsule, the human operators, and even the computer codes. "Relatively small things can make a big difference," Mark Herrmann told me, when we talked about the painstaking precision that's needed. A few percent error in the energy at the start of the laser pulse can degrade the conditions needed for fusion by as much as 50 percent.[12]

So far, NIF has not reached net energy gain. The best experiments have reached a fusion energy yield of 3 percent of the 1.8

sticks of dynamite's worth of energy in the laser beams. Given how fusion energy can scale up suddenly in the right conditions, this is far closer to JET than the difference in percentages alone might indicate, and the scientists here are front-runners in the race to build a star. NIF leads the pack on inertial confinement fusion and, some of the star builders here tell me, has the only machine that's up and running and could go all the way to net energy gain. This ambition is even reflected in the "ignition" part of the facility's name.

To learn more about what's stopping them from leapfrogging JET, I tack back to NIF's main building with Bruno. We head for the heart of the reactor, a catacomb of concentric shells of concrete and metal that house the business end of NIF. Inside the gap between the spherical aluminum target chamber and a bubble of thick concrete we can see metal gantries arrayed at all heights. These give access to the target chamber, which is neatly fed by tubes and cables from every direction.

Bruno points out the equipment that scientists use to see into a star. Every tube, box, and wire that isn't bringing a laser beam in is part of a diagnostic that tries to pick apart, often indirectly, what happened during each shot. Just as a doctor looks for symptoms to try to diagnose a disease, so scientists at NIF collect data on each implosion to diagnose what went right or wrong.

There are 250 people solely working on diagnostics here, whether that be on cameras that look into the chamber, fast laser pulses that probe the fusion plasma, neutron detectors, spectrometers, or the design of "diagnostic" targets that are modified to give firsthand information about conditions during an implosion.

One of those staffers is X-ray imaging diagnostician and group leader in physics Dr. Louisa Pickworth, whom I'd arranged to meet back in the visitors' center.

I had a strong memory of Louisa from my time at Imperial College London, where she'd spent hours in the bowels of the physics department working on the subterranean two-storey plasma machine known as MAGPIE. MAGPIE is a z-pinch, a machine that runs a large current through thin metal wires stretched vertically. The current breaks the wires down into a plasma and also creates a strong magnetic field. The combination of current and magnetic field pinches the created plasma into a dense vertical column. The "z" is because physicists often call the vertical direction the z direction.

MAGPIE can be used for lots of exciting science, for example re-creating in miniature the supersonic jets of plasma fired out of stars as they form.[13] Louisa researched new diagnostics sensitive enough to probe these experiments. Turning oil, metal, and electricity into scientific discovery using MAGPIE required both manual dexterity, to fashion targets and repair machinery, as well as mathematical rigor, to understand what was happening. She's now applying the same skills at NIF, a machine on a dramatically different scale.

"We haven't got fusion schemes working yet, and one of the most important things for us to do is to understand why," Louisa tells me over coffee. "We have many models and many computational codes which inform us how to put an experiment together and what we expect that experiment to give us, but then we do that experiment and nature gives us the real answer. And the job of the diagnostician is to listen to what nature is telling us."

Louisa first got interested in fusion back when she was sixteen, when she spent her summer acquiring work experience on JET. She was a diagnostician right from the start, developing an infrared camera to watch JET's magnetic confinement fusion plasmas. "My first experience of physics was working as a physicist at a big, expensive fusion facility," she says, "and to some degree I just fell in love with it."

Apart from the machinery, Louisa also enjoyed the international environment (something that both JET and NIF have) and the sheer complexity of coordinating so many people to try to make something amazing happen. But what she fell in love with most was trying to understand what's happening in extreme conditions.

"These are pretty challenging environments," she says, talking about the small size of the capsule during the implosion. "It's spherical," she adds, "which is a challenging geometry to image, and it emits a really wide range of radiation."

The diagnosticians look at all the types of light radiation that come streaming out of the fusion plasma. This includes optical light, from the capsule's periphery, and high-energy gamma rays, but Louisa's specialty is X-rays. From the pattern of X-rays that are emitted, she and her team have to work backward to find out how the implosion proceeded.

"Part of the diagnostic challenge is to try and take the information that's coming out and turn it into something useful!" she says.

One of the many difficulties she and her colleagues face is that the diagnostics on a real fusion shot must be indirect, and not interfere at all with the target. That's why they rely so much on sensors that work at a distance. That doesn't mean the tools are sophisticated; one of the workhorses is the pinhole camera, first developed in Ancient Greece. "It's literally a hole in a piece of material with a detector not dissimilar to film behind it," Louisa says. "It might tell us how round the imploded plasma is, how small it got, what sort of X-rays were there."

Those X-rays appear because electrons in the capsule's core are constantly being accelerated and decelerated as they encounter other charged particles. A charged particle rapidly changing direction creates an electromagnetic wave, in this case an X-ray, much like a speedboat pulling a tight turn creates a big wave in water. Because it happens as electrons screech around corners, this radiation is named bremsstrahlung, after the German words for braking, *bremsen*, and radiation, *Strahlung*. As you'll know if you've ever broken a bone, X-rays can penetrate material that regular light can't. Even though the plasmas produced at NIF are opaque to visible light, bremsstrahlung X-rays can pass through them. It's one of the ways for energy to escape from a fusion plasma that John Lawson put in his equation for net energy gain. Scientists like Louisa can use the escaping X-rays as a window on the experiment.

Part of Louisa's detective work is tracking down the instabilities that can ruin otherwise perfect implosions. Like magnetic confinement fusion, inertial confinement fusion is bedeviled by plasma instabilities. Those instabilities are a big reason why NIF hasn't gone further than a 3 percent yield of energy. "Nuckolls thought a kilojoule laser would do it," NIF's Jeff Wisoff told me. "When you start stacking in realities, Mother Nature has made this very hard. Mother Nature made actual instabilities much harder."

The most ubiquitous and feared instability among inertial confinement fusion star builders is the Rayleigh-Taylor instability. It happens when a less dense material pushes a more dense one, and its effect is to mix up the two materials involved. Like black coffee and milk, once mixed, the materials can't be unmixed. It may sound innocuous enough, but it's perhaps the single biggest issue for inertial fusion machines because it stops a hot, central core of fusion fuel from forming and taking hold. At NIF, it rips through the delicately arranged capsule layers.[14]

Recently, when I went to visit my old school in the UK's Peak District, we created our own Rayleigh-Taylor experiment with glasses of water and laminated sheets of paper. We filled the glasses to the brim, placed the laminate on top, and then turned the whole thing over. Try it at home. If you're really, really careful then you can remove the laminate and have an entirely full glass of water sitting, quite happily, upside down.

Instabilities need a tiny seed from which their exponential

growth can start, just as a pandemic starts with a single infected person. Without any imperfections in their interface, water and air can sit in a delicate, unstable balance. It was at this point, with two dozen glasses of water precariously looming over the classroom, that I told the students to give the water the tiniest of nudges with a finger or a pen tip and—almost instantly, as the instability grew, the water lost to gravity at every desk. We were mopping up the classroom for some time afterward—but the students had learned an important lesson about instabilities.

It's also the Rayleigh-Taylor instability that causes the very worst escapes of energy via bremsstrahlung radiation. "A flake of the hohlraum can fall off and get into the capsule," Mark Herrmann told me when we were discussing the latest challenges. This can create a jet, a finger of Rayleigh-Taylor growth, which pushes all of the capsule's layers deep into the fuel's hot core. That includes the outer layer of diamond-like carbon, which is thirty-six times more effective at radiating bremsstrahlung than hydrogen. Once carbon gets into the core, it aggressively cools it. "We can see it creates a *meteor* of X-rays," he added. Louisa Pickworth's team is tasked with tracking bits of the hohlraum, the gold, and the diamond-like carbon, to see how and why they get to where they have no business being. Although the Rayleigh-Taylor instability is the most acute, there are dozens of other ways for the perfect inertial confinement experiment to go wrong.

Given these truly formidable scientific challenges, you may wonder why inertial confinement fusion star builders have kept going. Just like star builders doing magnetic confinement fusion, they know that Lawson's equation means that net energy gain is possible in principle. But star builders pursuing inertial confinement fusion have another ace up their sleeve, one that the magnetic confinement fusion star builders don't: there's evidence that inertial confinement fusion can produce net energy gain not just in theory, but in practice. There's just one problem: that evidence remains top secret.

In the late 1970s and 1980s, a series of classified experiments, known as Halite and Centurion, that used nuclear explosives was conducted underground by the US national laboratories at Livermore and Los Alamos, respectively. We don't know much about them because they've remained only partially declassified because they were conducted as part of nuclear weapons programs. What can be guessed is that they involved indirectly driving an inertial confinement implosion. The driver, in this case, was a massive dose of X-rays from a conventional nuclear weapon rather than X-rays indirectly generated by a laser. However, the X-rays took up the rest of the process as if it were laser fusion. The US wasn't alone; the UK ran an independent version of the Halite-Centurion tests in 1982 that had exactly the same results. Even less is known publicly about these UK tests, apart from that they agreed with the US tests—and that one of the two designers was my PhD supervisor, Professor Steve Rose, now at the University of Oxford.

What we do know about these secret inertial fusion tests is that they showed "excellent performance, putting to rest fundamental questions about the basic feasibility of achieving high gain" in inertial confinement fusion. High gain, at least ten times energy out for energy in, is a basic requirement for a fusion power plant. Inertial fusion has been shown to work experimentally, albeit with a huge input of energy—much greater than what NIF's current laser can deliver. What we don't know is how close to the upper bound given by these experiments the laser energy needs to be before net energy gain can be achieved. It's exactly this question that NIF is now trying to answer.[15]

Inertial confinement fusion with lasers, and magnetic confinement fusion with Russian-style tokamaks, have both been around since the 1960s. Early star builder John Cockcroft, who first split the atom, thought it might be fifty years to a fusion power plant. That was in 1958. We know from John Lawson's theory that star power on Earth *can* work in theory. The leading magnetic fusion machine, JET, has come close. Inertial star builders think that secret experiments have already shown that their approach works in principle. But despite a succession of bigger and better machines, star builders working at these big laboratories still haven't managed to get to net energy gain. Maybe it's time for a different approach?

CHAPTER 7

THE NEW STAR BUILDERS

"Ideas alone have little value. An innovation's importance lies in its practical implementation."

—Werner von Siemens, cofounder of Siemens & Halske[1]

Most star builders are looking to the heavens for inspiration. Not Dr. Nick Hawker, CEO and CTO of First Light Fusion. He's thinking differently. He's looking under the sea instead.

The creature that set Hawker on the path to fusion is, by many accounts, the noisiest in the ocean: the pistol shrimp. Even though each one measures only a few centimeters in length, collectively they can drown out animals as big as whales. They are so noisy that, during the Second World War, US submarines hid from Japanese sonar among large colonies of them. The sound emanates from their claws. At a distance, mariners describe the clicking as resembling the crackling of burning tinder. Up close, the creature is much more deadly—at least, to other subaquatic denizens.

One of the pistol shrimp's two claws, the one that makes the noise, is about half the size of its body. The loud snap of a pistol shrimp is not, as scientists once thought, due to the sudden shutting of the large claw itself, but due to the bubble it makes as it closes. When that bubble of air pops, by imploding in on itself, it generates a shock wave that is heard as a CRACK. Pistol shrimp use the pressure of the shock wave to stun their prey before safely devouring them.[2]

The implosion of the bubble created by the pistol shrimp is so intense that it produces a tiny plasma that emits its own light. Long before humans mastered fire, there was an entire family of shrimp that could create plasmas of five thousand degrees under the ocean.

It's a little humbling.

Nick Hawker didn't have much interest in doing a PhD until the person who would become his supervisor offered him a position trying to re-create the pistol shrimp's plasma in a computer simulation. Nick's work in understanding this effect has led him from crustaceans to cooking up a star at his company, First Light Fusion. Nick explains that they wondered, "If we keep increasing the intensity of the shock wave, does the temperature keep going up or does it go unstable and stop working as most fusion things do?" The answer, the First Light Fusion team believes, is that these shock waves can reach fusion conditions.

Shock waves, like the sonic booms made by aircraft flying faster than the speed of sound, are special. Extreme ones can, in nanoseconds, increase the density and temperature of a material to the point that its atoms dissociate into plasma. They're often a key part of how inertial confinement fusion works, including at NIF. A pistol shrimp can create a bubble-collapse shock wave that Nick describes as "quite modest" at two kilometers (a little more than a mile) per second. That's the speed of the fastest ever crewed rocketplane. First Light matched that implosion speed with their first star machine, a gas gun that used a small chemical explosion to propel a solid projectile. They're now on their third machine.

As I walk around First Light Fusion's laboratory, it's hard not to be impressed by the clean, tight ship they run. When I enter the building, I find a reception area that is all start-up, complete with plush chairs and bright colors. Whiteboards, wiped clean for my visit, adorn the walls. There's a fancy coffee machine in the kitchen. In a nice touch for a lab that must be obsessed with diagnosing temperatures, every water jug has a little thermometer attached to it. And the areas where the experiments are done are absolutely spotless.

How First Light Fusion plans to build a star is to accelerate a small piece of material—they wouldn't tell me what but probably a slug of metal not much bigger than a coin—into a target that contains deuterium and tritium. When the slug smashes into the target, it produces a shock wave with enough energy to create a high-temperature, high-density plasma. They calculate that they need a collapse of fifty kilometers (approximately thirty miles) per second for fusion, but—and this is their top secret innovation—the way their shock wave interacts with their target multiplies the collapse velocity many times over. Their approach is based on inertial confinement and so has similarities to laser fusion, except that they've replaced the laser pulse with a "pulse" of solid matter.

First Light Fusion's latest step on the way to a full-blown star machine is an electromagnetic rail gun named Machine 3. It has been lovingly assembled as a six-limbed fourteen-mega-amp monster. It charges in fifty seconds and deposits all of its energy into the projectile in just two microseconds. Each shot uses 2.5 megajoules of energy. It is electromagnetic because chemical processes can't go fast enough. The rail gun uses a current in one direction (along one rail, through the projectile, and back by the other) and magnetic fields (in a cylinder around each rail) to force a projectile to accelerate in the third direction.

In principle, this electromagnetic rail gun can propel objects to speeds of twenty to thirty kilometers per second, but so far, First Light can only be sure they've hit fifteen. A satellite reaches seven kilometers per second during reentry from Earth orbit. I know this because First Light Fusion's head of numerical physics, directing ten people and a supercomputer with two thousand CPUs, previously modeled satellite reentries for a living. Even at this speed, the shock wave that builds up in front of the satellite is strong enough to rip apart atoms into plasma. At tens of kilometers per second, First Light is well into the plasma physics regime.

Even though I've visited star builders at top secret government laboratories, at a site handling nuclear materials, and at other start-ups, First Light Fusion is by far the most secretive. No mobile phones are allowed beyond the reception area, a restriction that didn't even apply at Lawrence Livermore National Laboratory. Most visitors who see the inside of First Light Fusion's offices are asked to sign a nondisclosure agreement. When I ask about all this secrecy, they explain that the target technology—what turns their shock wave into a fusion initiator—is a trade secret. Hawker tells me that patents aren't much good if you're a small fry because patent battles are always won by the bigger fish in court. But no one can copy a trade secret as long as demonstrable effort has been put into protecting it.

It's interesting that First Light is so averse to patents. The assumption made by some fusion start-up skeptics is that the real business plan of private sector star builders is to either generate a few high-tech patents to sell off or to be bought out by another firm. Flogging patents, whether to investors or to other firms, is not a new trick; there was a significant boom in the 1690s as firms piling into the sunken treasure industry sought to establish their credentials (Daniel Defoe was among those caught out by the "patent-mongers").[3] Selling up is a common strategy for start-ups, and when it works well, both investors and employees can come out happy. But, like the pistol shrimp, First Light has no intention of being swallowed by a bigger fish.

"For us, getting bought is actually plan B or C," Gianluca

Pisanello, the chief operating officer, tells me. The business plan is to retain control over the rights to the secret fusion targets, but to let big industry build the power plants. "At the end of this process, we will have two fundamental pieces of intellectual property in our pocket: one is the target, one is the driver. We will make money by selling the targets."

Like Nick Hawker, Gianluca Pisanello is typical of the new wave of star builders. He's an engineer by training. His dream was to be a Formula 1 racing engineer. After studying electronic engineering, he made his dream come true, working for Toyota and other teams, and eventually becoming chief engineer for a team of sixty people. When I commented on the astounding cleanliness of the laboratory, he flushed and said that he tried to bring his Formula 1 standards to fusion. Gianluca's job as chief engineer in the racing business was to optimize every aspect of car plus driver to gain fractions of a second. It was extreme engineering. Or so he thought, until he became a star builder.

One day Gianluca received a call from a headhunter who told him about First Light Fusion. Even though it seemed like a long shot, and he didn't know much about fusion at the time, he couldn't quite bring himself to say no. Eventually, he said yes.

"I realized I didn't care if we succeeded," he told me. "What I wanted was to be part of this journey, and if this happens, and there was an opportunity to be involved, and you didn't take it . . . well!"

Every star-building start-up claims it has what it takes, that it's different from the competition, special. First Light Fusion's story is about trying to use the most tried and tested technologies wherever possible. This way, the innovations, and the risk, will be concentrated in the targets. By buying almost everything else off the shelf, Nick Hawker and his colleagues reduce the overall risk of what they're doing. I can't help but admire the clarity and pragmatism of the vision.

While Gianluca Pisanello thinks the big laboratory competitors, JET and NIF, could get energy gain, he and First Light believe their approach sidesteps the big fusion engineering challenges that will make it difficult to build a first-generation power plant.

First Light Fusion's vision for a power-producing star machine involves one target at a time being dropped into a chamber, followed by a much faster projectile that catches up to it. When the two collide, fusion reactions will be triggered. Surrounding the fusing plasma will be a cylindrical wall of liquid lithium—imagine standing inside a circular waterfall, but instead of water, it's metal. The liquid lithium will absorb the neutrons, making precious tritium that can be used as fuel. Heat energy carried by the neutrons that gets into the lithium will be exchanged into another medium like water. Ultimately, the water will be turned to steam to drive a turbine. The whole process will repeat somewhere between every five or every forty seconds.

Before coronavirus struck, First Light Fusion had a plan to perform a net-energy-gain experiment by 2024, and they say they're about to reach the temperatures where fusion reactions first become detectable. But what both Gianluca and Nick are keen to stress is that they're not in the net-energy-gain business: they're in the power business.[4]

"It's the world's largest problem," Nick tells me, referring to achieving fusion. "Most scientists are working on the physics, which is a bit of the problem but not the whole problem. Demonstrating gain is not valuable. Heat and light are valuable."

One challenge that I level at him is that they're making their most important innovations secret, but science moves forward more quickly when it's done out in the open, where ideas can be critiqued and improved upon.

"Sure," he says, "but does technology move forward more quickly?"

Engineers in Charge

Thanks to decades of publicly funded research, fusion is increasingly a technological challenge rather than a scientific one (though big scientific challenges remain). To take it to its next phase, new types of talent will be required—those who can convert scientific ideas into working technologies. Enter the engineers.

Engineers like Nick Hawker and Jonathan Carling, CEO of Tokamak Energy. The latter told me that "things like the steam engine and internal combustion engine were invented long before anyone understood how they worked. Once it kind of works, engineers will take over." And so they are.

Just twenty miles south of First Light Fusion's office, past Oxford's dreaming spires, is the industrial park that houses Tokamak Energy, a fusion start-up that uses a radical tokamak design. Until very recently, Tokamak Energy's building sat in the shadow of the great cooling towers of Didcot Power Station. For years Didcot burned oil, coal, and gas, and pretty much dominated the landscape. That is, until the towers were brought down by controlled demolition. How appropriate that this part of Oxfordshire, which is also home to JET, should see this emblem of fossil fuels fall to the ground and so many innovative fusion schemes appear.

Tokamak Energy has raised more than £117 million (about $158 million) in private investment and is looking for as much as £700 million ($936 million) for a future phase. Its star builders are nothing if not ambitious. The former CEO, now executive vice chairman, Dr. David Kingham, brought Jonathan Carling in because of the need to transition from fusion-in-principle to fusion-in-practice. David is a theoretical physicist by background, but for most of his life he has been involved in the UK's high-tech start-up scene, previously running business accelerators that have helped launch several thousand firms.

"Fusion was always seen as too hard," he tells me in Tokamak Energy's meeting room, "the preserve of big government labs. That was very much a dominant view of the world until just a couple of years ago." He's excited about what the private sector can do for fusion. He sees the relationship between private fusion ventures and the big laboratories—like Culham—as being analogous to the relationship between SpaceX and NASA. He's not alone—everyone I speak to in fusion start-ups keeps coming back to that analogy with space.

Tokamak Energy has taken the usual tokamak plasma shape and squished it so that it looks less like a doughnut and more like a cored apple. The resulting machine is called a spherical tokamak. The rationale behind this is really simple: magnetic fields dissipate over distance, so when the plasma is brought in closer to the core (which generates the toroidal magnetic field), less magnetic field is needed for the same amount of confinement. It also creates a more compact machine, which—given the increasing sizes of each generation of tokamaks and their high costs—would make magnetic fusion power more economical. David enthusiastically explains that there's a lot of academic work showing that spherical tokamaks will be able to deliver fusion more quickly, with smaller, cheaper machines.[5]

But that's not the only innovation that Tokamak Energy is introducing. It also plans to substantially increase the strength of the magnetic fields using high-temperature superconducting magnets. "High-temperature" here is a fairly silly term—the highest temperature these magnets operate at is twenty degrees above absolute zero. The old "low-temperature" ones only worked at two degrees Kelvin, so it's all relative.

Magnetic resonance imaging, or MRI, machines in hospitals

use superconducting technology and need to stay below around ten degrees Kelvin. They're about the highest magnetic fields that you might experience in everyday life, capable of generating between one and three Teslas of magnetic field strength. This is similar to what Tokamak Energy is aiming for in its machine. Due to the superconductors and the plasma being so close, Tokamak Energy's spherical devices will have one of the wildest temperature gradients in the solar system. The tokamak plasma has to be at least 100 million degrees while, a meter away, there is a superconducting magnet at 20 degrees above absolute zero.

Magnetic fields provide most of the confinement in tokamaks: the stronger the better. The fusion energy ramps up aggressively, so that an increase in the magnetic field by a factor of two means producing sixteen times as much energy per second. But it takes a lot of energy to push current—or a stream of electrons—through the tokamak's core, usually a cooled copper wire, to create the toroidal field. When superconducting materials are used, the electrons encounter almost no resistance.[6]

There's just one problem.

"Low-temperature superconductors are on a hair trigger and can be easily quenched by any amount of energy," David Kingham tells me. *Quenching* is a sudden and explosive increase in electrical resistance of the conductor as it ceases to be super. A huge bang issues forth as electromagnetic energy is rapidly converted to heat, which boils off the coolant. Modern MRI machines are carefully designed to safely quench at the touch of a button, but this is no good for a power plant that needs to remain on as much as possible. Severe quenching could twist metal and permanently damage a reactor chamber.

The researchers at Tokamak Energy aren't alone in thinking that the future will be an advanced type of tokamak. Commonwealth Fusion Systems, born out of a well-respected fusion science program at MIT, will also use superconducting technology in a new tokamak design. They believe their advanced setup will allow them to get to a fifth of the fusion power of the biggest tokamak ever designed while only having one-sixty-fifth of its volume. Commonwealth's scientists have recently published research in academic journals that suggests their tokamak could comfortably deliver twice as much power out as is put in.[7] Clearly, some investors agree because Commonwealth has received $200 million in funding. In 2018, the company's CEO said, "We think we have the science, speed, and scale to put carbon-free fusion power on the grid in 15 years."[8]

Tokamak Energy is aiming to demonstrate in principle that its machine can reach the point of a net gain in energy by 2022. This is not, strictly speaking, "more energy out than in" from fusion reactions, because they don't want to have to work with tritium. Instead, they're going to put the much more plentiful, and easier to handle, deuterium into their reactor and show that they can get to the *conditions* that Lawson's equation says are needed to generate more energy than put in. Jonathan Carling, like Nick Hawker, explains that gain is not the point though:"The main thing we're trying to prove is that our modeling is correct. Whether it's just shy or just over [net energy gain] is not the important thing for us, it's whether it's on the curve we predict."

The hope is that demonstrating that their modeling is correct will be enough to convince investors to fund a machine that can—in theory—produce way more than 100 percent of the energy put in using deuterium and tritium. Other fusion start-ups have similar plans.

Like First Light Fusion, Tokamak Energy is much more interested in power production than energy breakeven. "Achieving a Q of one is a scientific goal," Jonathan Carling continues, "but it's nowhere near enough to produce commercial energy, which requires a

Q [in the region] of tens." As mentioned previously, Q is the ratio of fusion power out to heating power in. His strong view is that unless other star builders have a credible plan to get to factors of twenty or thirty more power out than they put in, then they're in the science game and not the fusion energy game. And he puts JET, the conventional tokamak at Culham, firmly in the science category.

Promises

The promise of gain (or gain conditions) by 2022 and the mid-2020s by Tokamak Energy and First Light Fusion, respectively, may seem overly optimistic to star builders in government laboratories. But these punishing timescales aren't unusual among the new wave of private sector star builders, such as General Fusion, LPP Fusion, Lockheed Martin, TAE Technologies, HyperJet Fusion Corporation, MIFTI, Proton Scientific, Helion Energy, Commonwealth Fusion Systems, Renaissance Fusion, Zap Energy, HB11-Energy, Pulsar Fusion, and the list goes on. There are now more than twenty-five private sector fusion firms. Most are promising to deliver energy from nuclear fusion reactions in years rather than decades. Commonwealth Fusion Systems says it will achieve net energy gain by 2025 *and* a pilot power plant by 2033. The deadly Covid-19 pandemic may slow these timescales, but the intention is clear: fusion sooner rather than later.[9]

There's fierce competition between the star-building start-ups. It's hard not to imagine the race to demonstrate gain—and, even more so, to build a prototype power plant—being a winner-take-all contest. The victor will enjoy an influx of investors looking to ride the new nuclear wave. The losers may be able to bask briefly in reflected glory, but unless they can also demonstrate gain quickly, fusion dollars are likely to go to the firm that's already made it happen.

Arguably, being exposed to market forces hands them all a big advantage. Whichever firms slip behind are probably going out of business, which must provide a sharp motivation to stay ahead. Without the bureaucracy of thousands of staff or public sector procurement rules, these start-ups are able to be much more agile than government laboratories. "I have no new physics for the world," reality TV show *Made in Chelsea*'s Richard Dinan, CEO of Pulsar Fusion, has said. "My trick is that I can build technology quickly and cheaply."[10] If a line of inquiry doesn't work, they can shut it down, fire the staff, and refocus their resources on more promising avenues. Nick Hawker told me that he was glad that they'd gone private, because, otherwise, they'd be tangled in endless arguments about where to build their gain-targeting star machine. Instead, he said, "We can put it wherever we like: we're a private company."

They have other advantages too—start-ups pay higher wages than universities and government laboratories outside of the USA, so they can attract top talent. Two of my former colleagues at Imperial College left to work for a start-up because the pay and job security are better than in academia. "They had the money," a scientist whose contract with the University of Oxford had run out told me.

Some of the scientists leading this new entrepreneurial wave are undoubtedly as eccentric as those from fusion's not-always-stellar history, but their fresh approaches are attracting eye-watering amounts of both private and public investment. Recognizing the benefits of competition and innovation in fusion, the US Department of Energy in 2020 released $61 million in funding.[11] Zap Energy and Commonwealth Fusion Systems are among the beneficiaries. These private firms are proposing to use millions of dollars, and some crazy ideas, to do what billions of dollars, and decades of scientific investigation, have been unable to.

Of course, everyone thinks that their firm is closest to net energy gain, that only *they* have a viable plan for fusion power, and that it's the *other* star builders' plans that won't work out. When I tell Jonathan Carling this, he smiles. "There are lots of respected ventures out there, but we do believe that we're the only business that's got a commercially viable technology and an operational plan to get there."

When I press him on why his firm is the special one when everyone says that it's their firm, he is a little more candid about the competition: "Someone's going to do it a lot faster—we're pretty convinced it's going to be us because we've got a combination of agility, the right people, the right money, and we're basing it on known tokamak science. We're not trying to invent, you know, I've got a new thing with three hundred pistons and whatever, or I've got a shrimp, or all these other things . . ."

No prizes for guessing which crustacean-inspired star builder he's referring to. Jonathan's mention of pistons appears to be a reference to General Fusion, a firm based in Canada that has been operating for almost twenty years. It was founded by Michel Laberge, who quit his job as a senior laser printing engineer to follow his fusion dream. In 2019, General Fusion announced that they'd raised a further $65 million, taking their total to at least $200 million. Microsoft, the Canadian government, and Amazon's Jeff Bezos are all said to have put money in.[12]

General Fusion plans to build a star by confining a ball of plasma with liquid metal walls that are themselves confined within a reactor chamber. The idea is that the liquid metal will be pushed by steam-driven pistons until it compresses the plasma to fusion conditions. The neutrons emitted will be captured by the liquid walls. It's not clear (to me) how the pistons will act quickly enough or forcefully enough to take the plasma to the extreme conditions that fusion requires. As those working at NIF know all too well, a heavy fluid (liquid metal) pushing on a light one (plasma) is a nightmare for Rayleigh-Taylor instabilities that could mix the liquid metal through the plasma. At NIF, compression is provided by shock waves that are themselves a result of a laser pulse that, naturally, travels at the speed of light. In First Light Fusion's machine, an electromagnetic rail gun accelerates an object to tens of kilometers per second. It's hard to imagine steam producing the speed or strength of compression needed to hit stellar conditions. There'd also be a lot of heat energy leaking through the metal-plasma interface. It's a steampunk fusion scheme that seems to defy all the usual wisdom on confining plasmas. Of course, much of the technology is secret, so no one outside the firm really knows how it works, or how it will work. None of this has put off the big name investors.[13]

What I hear from a lot of star builders—and what Jonathan Carling's critique of what some might describe as "wacky new schemes" is getting at—is that all fusion schemes have a tendency to go awry in unexpected ways. In new fusion schemes, problems are equally likely to manifest, it's just that they haven't yet been discovered. This is why Nick Hawker and First Light Fusion's philosophy is to use off-the-shelf technology for everything except for their targets. It's why Jonathan Carling and Tokamak Energy are putting their faith in spherical tokamaks, a variation on an existing technology. When start-ups pursue fusion schemes that are less well tested at large scales and promise that they can reach net energy gain in a few years, it does tend to raise eyebrows. But investors seem to have an appetite for even these schemes.

In fact, the best funded of all the start-ups, TAE Technologies, is pursuing a very risky approach, albeit one with commensurately bigger rewards. "Start-up" hardly seems appropriate for a firm that has been in the business of fusion since the 1990s. TAE has a device that fires the plasmas produced by two pinches into each other to create a ring of plasma that is briefly bound in magnetic fields, like a smoke ring in air. It's an unusual scheme but one that is also being pursued by Helion Energy, a firm backed by PayPal founder Peter Thiel.

The big risk that TAE Technologies is taking is in the fusion re-action it's trying to produce. While almost every other fusion firm is focusing on the easier deuterium and tritium reaction, TAE is trying to initiate reactions between the regular form of hydrogen (a proton) and a common isotope of boron. The advantage of this reaction is that it doesn't produce neutrons, just helium nuclei and energy. Removing neutrons from fusion is very attractive: it means close to zero radioactivity, no need for thick shielding, no tritium breeding, and no degradation of the reactor chamber over time. It's much more commercially appealing. There's a catch though (there always is): the temperatures required for this neutron-free fusion reaction are at least *ten times* what is required for deuterium-tritium fusion. Even at that temperature, the number of fusion reactions is fifty times fewer.

TAE is backed by Google, Goldman Sachs, and the Russian government (Microsoft's cofounder Paul Allen was also an investor before he passed away), and has raised a whopping $700 million. Some star builders think that we won't see a working proton-boron fusion reactor for decades, at the earliest. TAE has reported significant progress on its machine, but they need to improve its performance ten thousand times over just to reach the conditions needed for deuterium-tritium fusion—and proton-boron fusion is even further out of reach.

With such backers, science may not be the only challenge. One fusion start-up insider I spoke to raised the possibility that the ag-gressive hiring of PhDs and other highly talented employees meant that TAE might divert their energies into other, more immediately profitable activities. The company has already spun out one firm, TAE Life Sciences, which produces physics-based cancer treat-ments. Yet TAE is making big fusion promises: in 2019, the CEO, Michl Binderbauer, claimed in an interview that his team would be producing "energy from fusion in two years." Subsequently, he backtracked to "a small number of years." Another comment from 2019 was even bolder: "We're talking commercialization coming in the next five years for this technology."[14]

There's one star-building firm that has even more money to draw on than TAE Technologies: Lockheed Martin, whose market capitalization is just over $100 billion. Its size and scale give it the ability to fund fusion even beyond most nations. Lockheed Martin's fusion team is pursuing yet another approach, which combines magnetic mirrors—which can be used to bounce the plasma around—and a magnetic cusp, which creates a hollow in magnetic field lines that can trap charged particles. The company's fusion program has been going on behind closed doors since 2010, and in 2014 they claimed that they could "design, build, and test the new compact fusion reactor in less than a year" and have a prototype truck-size power plant in five years. You may have missed the news about Lockheed's solving the problem of controlled fusion in 2015: I certainly did.[15]

Ian Chapman's predecessor as CEO of the UK Atomic Energy Authority, Sir Steve Cowley, has expressed surprise "that a company like this would release something that doesn't have much context. Normally, if someone says they're doing well in fusion, they would quote some data," which seems a polite way of saying that these claims are likely significantly exaggerated.

Lockheed Martin isn't the only firm to have over-promised and under-delivered on fusion. Steampunk fusion outfit General Fusion claimed in 2009 that they'd build a working power plant within a decade. They haven't, not that it has slowed them down a jot. A decade on, although there's no power plant, they do have a new CEO and a big machine. That new CEO, Christofer Mowry, is now aiming at 2023.[16]

Another brimming-with-confidence fusion player is LPP Fusion, formerly Lawrenceville Plasma Physics. LPP's CEO, Eric

Lerner, is a controversial character.[17] He has written a book detailing why he believes the Big Bang never happened, directly challenging the overwhelming scientific consensus. His firm is trying to do fusion using a dense plasma focus, a pulsed power device that has been around since the 1950s. It creates a pinch effect not in a line or ring, but in a blob of plasma above one end of a cylindrically shaped conductor. LPP's device is very small, about one centimeter long. The small size, and therefore low cost, of dense plasma focuses makes them perfect sources of high-energy neutrons, which is what they're usually used for.

Star builders whom I spoke to pointed out that dense plasma focuses of the kind LPP is using almost exclusively produce beam-target fusion—useful in some contexts, but not for power production. Beam-target fusion is similar to what Rutherford, Oliphant, and Harteck did when they discovered nuclear fusion; it smashes a few particles together, and so produces neutrons, but it's not scalable to a power plant. The difference is in the temperature: going back to my analogy of hot plasma being like children running around a playground and colliding, beam-target fusion is like one child briefly running through a playground of children pretending to be statues. You might get *some* fusion, as the moving child collides with a stationary one, but you won't get significant energy that way. For this reason, fusion with hot temperatures—the kind that most star builders are doing—is called "thermonuclear" fusion as opposed to beam-target fusion. The difference between beam-target fusion and thermonuclear fusion is akin to the difference between fool's gold and real gold. Getting the distinction wrong has been the downfall of many fusion schemes, even those run by scientists at the absolute top of their game. To make it more challenging LPP Fusion is, like TAE, going straight for the more tricky neutron-free fusion reaction between hydrogen and boron.[18]

So what promises has LPP's controversial CEO Eric Lerner made?

In 2014, he claimed that LPP was closer to affordable, unlimited, and ultra-clean energy than any other fusion outfit in the world. At that time, LPP had already secured more than $3 million in funding. He told Fortune.com that, with the financing, he'd be licensing the mass production of small fusion machines by now. He isn't.

LPP went on to use crowdfunding platform IndieGoGo to convince the public to pledge bits of cash to help it build a working device. Unusually for a crowdfunding campaign, even if the targeted amount of $216,000 wasn't reached, any pledged money went to LPP Fusion anyway, and the target was "flexible" so that it could move during the campaign. The company raised more than $190,000. An article published at the time claimed that LPP would have a working device by 2020, and yet, if it has one, the world's media have been oddly quiet about it. As of 2020, LPP Fusion had raised another $600,000 through equity crowdfunding site We-funder. The accompanying video claims LPP is second only to JET in the race to net energy gain, but publicly released data suggest it's getting at least one hundred times less "energy out for energy in" than Lawrence Livermore's NIF. To its credit, LPP is at least publishing data on the progress it's making—something that can't be said of all fusion start-ups.[19]

Extraordinary claims require extraordinary evidence. Although almost all fusion start-ups have in common smart-looking home pages with animated videos of what their first-generation power plant will look like, the worry is that techno-futuristic pablum and a flashy website may be *all* that some fusion start-ups have to offer. For critics of some of the claims being made, there's plenty of grist for the mill.

A good number of the start-ups are sending their scientists to conferences to present their work and also publishing research re-sults in peer-reviewed academic journals (though they don't pub-lish *all* of their secrets, of course). There's a degree of transparency and openness about what they're doing. From others, very few details have emerged, so it's hard to judge their ambitious claims of how quickly they'll achieve net energy gain or the more ambitious goal of commercialization. Investors and scientists alike need transparency to judge just how far firms are toward net energy gain.

The risk seems great that one or more of the start-ups will blow up due to overly optimistic or, even worse, downright misleading claims. I spoke to one individual, heavily involved in the fusion industry, who was frank about a small number of fusion firms' schemes being bunkum, but they attributed much of it to an inability to give up on a dream rather than intentional manipulation of investors. Fusion is complex, and there's a risk of investors or even just ordinary people being hoodwinked. After all, fusion scientists themselves have got it wrong many times—the British scientific establishment, led by early star builder John Cockcroft, was humiliated when its "ZETA" machine was found to be doing beam-target, rather than thermonuclear, fusion in 1958. This was an unintentional but preventable mistake—the Soviets and the Americans never really believed the results. A more serious incident, due to blatantly bad science, occurred in the 1980s when the fusion community was briefly fooled by so-called "cold fusion." Instead of using high temperatures and plasma, cold fusion attempted to combine deuterium and tritium at room temperature using a catalyst. Initially, the results seemed so promising that Edward Teller even called up the discoverers to congratulate them. But cold fusion was bunkum. Apart from professional investors and members of the public losing their money, big promises that aren't delivered on can poison the well for both government laboratories and legitimate private sector fusion start-ups.

I wondered whether scientists pursuing more conventional fusion machines, who tend not to make such bold promises, would be troubled by some of the wilder claims. To my surprise, many of the scientists pursuing more accepted approaches were approving of their edgier competitors—or so it initially seemed.

"Fusion needs innovation, it needs new ideas. I'm not going to discourage anyone. Nothing would make me happier than if we could do fusion today," Dr. Mark Herrmann, NIF's director, told me. Ian Chapman, who leads the UK Atomic Energy Authority, felt the same: "The simple answer is that the more investment, public or private, in fusion, and the more smart people working on fusion, the quicker we'll get a solution." He paused for a moment before adding, more positively, "It's a good sign that the market is showing appetite for fusion."

Dr. Louisa Pickworth, the leader of a group of NIF diagnosticians, also began on a positive note. "Personally, I think the more the merrier," she told me, "there are probably many ways to cut this."

I pressed further, and Louisa admitted that she does worry about what will happen if investors in fusion start-ups don't receive a return and it puts people off in the long run. And although Ian Chapman welcomes the competition, he doesn't actually think they've got what it takes: "My view is that the biggest private sector fusion company in the world is two hundred people, and they are no way capable of designing a full integrated system, and that is the biggest challenge of fusion."

Ian Chapman is also firmly of the view that it will take billions to scale fusion into a bona fide power source, and that the private sector just won't be willing to take that risk. Eventually, he admitted that the potential for broken promises from the private sector could be a problem.

"Yeah, look, fusion has really suffered for fifty or sixty years now from ZETA and from overpromising . . . Private companies hyping and over-exaggerating does run the risk of another false promise, another non-delivery from the fusion community," he told me, adding that the public sector had probably been too conservative about its own successes.

"I never want to discount that someone might come up with a smart idea," NIF's in-house astronaut Dr. Jeff Wisoff told me, "but people have been thinking about this for many years, and I'm fairly skeptical about a garage-size machine."

Even Professor Sibylle Günter, normally apt to give a balanced view, had strong words. She has no truck with start-ups that promise shareholders a quick return. "Some of them hide details of the scientific concept and results from the community behind a claim of commercial secrets, and thereby escape any competent criticism which could probably damage their standing with the shareholders." And, she added, scathingly, that the "promised timescales [are] totally unrealistic."

Günter also said that even the best start-ups investigating alternatives to tokamaks that still use magnetic confinement are "operating in plasma regimes that correspond to tokamak performance several decades ago," or are having problems that tokamaks never had. Her sharpest criticism, though, is for start-ups pursuing fusion reactions that aren't based on deuterium and tritium. "For the moment," she told me, "and I'm talking probably for a one-hundred-year period—they're completely unrealistic." Such a gap between rhetoric and reality may threaten all fusion efforts.

What do the fusion start-ups themselves think of the risk that one of them breaks their promises in a very public way? Tokamak Energy's David Kingham thought the benefits far outweighed the risks. "Private sector fusion grows the pie," he told me. But that doesn't mean he didn't have any concerns: "There's a risk of a cold fusion–type idea emerging, and you do see that from time to time. Because fusion is exciting and complicated there's a risk of insubstantial or fraudulent ventures emerging. There are one or two particular firms who haven't been allowed to join the fusion industry association."

David Kingham wouldn't tell me which firms. He doesn't have concerns about Tokamak Energy's sponsors being spooked, because "our investors are sophisticated." Nick Hawker tells me exactly the same thing. "If there was a high-profile failure that could discourage some investors, yes," he admits, "but we're fortunate because our investors get it and have been through bubbles before."

The strength of competition is such that they all think they're the only ones who will deliver fusion on time and well below the costs that governments are shelling out for.

Governments Are Behaving Like Start-ups to Stay in the Race

Don't count out the government laboratories yet though—they're adapting too. One of the early designs for a magnetic confinement fusion reactor that was out of favor for years, the stellarator, is now back in the race for fusion thanks to a project run by the Max Planck Institute for Plasma Physics, where Professor Sibylle Günter is the scientific director. The staff there has built the world's most advanced stellarator, Wendelstein 7-X (also known as W7X).

Stellarators solve the problem of particles in the plasma drifting out of confinement with a twist, literally. Instead of twisting the particle orbits around the inside of the tube with one externally applied magnetic field and another field generated by a current in the plasma, stellarators use a twisted combination of externally applied magnetic fields to guide the plasma inside the tube into doing helical twists. The magnetic field twisting is delicately balanced to cancel out the drift of particles toward one wall or another.

"The stellarator concept seemed doomed for a long time," Sibylle told me, "because experiments showed that particles, and thus energy, got lost too fast." But, she says, a better theoretical understanding of plasmas led them to realize there were conditions that wouldn't suffer this loss of confinement, and, in principle, they could build a device to those specs. However, creating the perfect, twisting magnetic fields—and fitting in the energy-hungry coils that generate them—is a tall order in practice, even if it's possible in theory. To make it work requires that the magnetic field be so perfect that the deviations are no bigger than one part in one hundred thousand.[20]

Stellarators are back in steel and concrete because two technologies brought the seemingly impossible requirements into reach. The rapid development of supercomputers has meant that star builders are able to design and simulate reactors with just the right twists and turns in the magnetic field to keep particles on track. The second technology is superconductivity, which means it's possible to create enormous magnetic fields without using too much electrical power.

Sibylle told me more about the design of the machine, which was completed in 2015. "Wendelstein 7-X is the first optimized stellarator of sufficient size to demonstrate that stellarators can achieve confinement properties similar to tokamaks," she says. And the confinement properties of tokamaks are close to what is needed for net energy gain. Unlike in tokamaks, the magnetic confinement is solely provided by external coils, with no plasma current. Fifty superconducting niobium-titanium coils, each 3.5 meters (approximately 11.5 feet) high and cooled by liquid helium, wrap around the helically twisting reactor torus to provide this field. While it's not what you'd call a compact machine, it's not huge either, measuring about 16 meters (approximately 52.5 feet) across. Sibylle says that stellarators have the potential to be stable and easy to control— not sentiments that you hear in relation to tokamaks.

This small(ish) star machine is more what you would expect

from a fusion start-up than a collaboration between the German government, the EU, and the US national laboratories. Within just a few years of operation, it had smashed through the previous stellarator record for fusion conditions with a combination of temperature, density, and confinement time that is 6 percent of what Lawson's famous equation predicts is needed for net energy gain. This is impressive stuff for a technology that was largely abandoned by the world following the then USSR's success with the tokamak (the then best stellarator in the US, the Model C, was ignominiously converted into a tokamak).[21]

Although it's a government-run machine, W7X has achieved successes that show why private fusion can't be counted out. New technologies allow for schemes that were once thought to be forever out of the running to be competitive again. W7X also demonstrates how smaller machines can catch up on decades' worth of progress.

Stellarators aren't the only machines that have received renewed attention from government laboratories. The fusion machine that Tokamak Energy is using, the stouter and more apple-shaped version of a tokamak, called a spherical tokamak, originated in government fusion programs and is suddenly back in fashion with state-supported scientists. I asked Ian Chapman, the scientist-cum-civil-servant with a politely frank manner and an unwavering belief in the good that fusion can do, what he would say as CEO of the UK Atomic Energy Authority if the EU or the British government said, "We want to accelerate fusion and we're going to give you the money: what will you do?"

"What we'd really like to do," he told me, "is explore the spherical tokamak as a reactor path and build a machine on JET scale that produces net electricity. JET was about £2 billion in today's money and took four years. If we built a spherical tokamak on that scale, it would cost more than £2 billion, but not twenty, and we could build it in less than a decade [once designed] . . . It would be exciting to try a smaller, cheaper, more compact reactor."

This is almost exactly what Tokamak Energy is trying to do, though they think they can do it with "just" £700 million (approximately $930 million). Spherical tokamaks, which give more bang for the buck in terms of magnetic field strength, simply weren't around when JET was first conceived. Ian tells me that the low-risk, low-commitment funding of fusion followed by governments since the early 1990s has necessarily meant that star builders have stuck with conventional tokamaks, partly because they're tried and tested. However, Culham has already cut its teeth on a small, experimental spherical tokamak that can reach 10 million degrees Celsius (18 million degrees Fahrenheit), called MAST Upgrade. Ian favors spherical tokamaks as a second strand now because they have a clear confinement advantage over the conventional machines. That said, they come with more risks—no one has ever built one at a scale even close to being relevant for power production. And precisely because they're more compact, all of the difficulties of dissipating the heat from fusion are exacerbated, calling for more complex heat exhaust systems.[22]

A few weeks after I visited Ian in his Portakabin-like office at the Culham Centre for Fusion Energy, the British prime minister had exactly the same conversation with him. Ian Chapman stayed true to his answer, and now there's £220 million ($396 million) to fund the initial design (not the construction) of what is being called the Spherical Tokamak for Energy Production, or STEP. It'll be just ten meters (approximately thirty-three feet) in diameter, a little bigger than JET, but—if built—would be designed to safely go beyond net energy gain and deliver around one hundred megawatts of power to the grid. It's not clear if Ian's having such a well-practiced answer when the PM asked his question contributed to his funding success.[23]

Another new government scheme with promise is MagLIF,

short for magnetized liner inertial fusion. Sandia National Laboratory sits in the outer suburbs of Albuquerque, New Mexico. Like Livermore, it's part of the USA's National Nuclear Security Administration. While Livermore has become the world's center of excellence in high-energy lasers, Sandia stuck with pinches similar to those first tried out as long ago as the 1940s.

Like Livermore, Sandia uses its biggest machine—"Z"—for a combination of open science and secret experiments related to nuclear weapons. It's the machine that Omar Hurricane, NIF's chief scientist, learned his craft on. It's the type of machine that Louisa Pickworth developed her diagnostic skills on before turning them to the implosions on NIF. The name Z derives from the machine's so-called Z-pinch. Instead of pinching plasma around a ring, Z pinches plasma in a vertical column.

Scientists at Sandia had the idea that they could use Z's capabilities for fusion. They take an empty beryllium can, which looks like a miniature baked beans can at just 1 centimeter high and 6 millimeters (less than an inch) in diameter, and fill it with deuterium and tritium. Then, they hook it up to the middle of Z and whack an absolutely enormous magnetic field—ten to thirty Tesla—through the can lengthwise. This is equivalent to at least seven MRI machines, or, putting it in Ig Nobel Prize terms, it's enough to levitate a frog.* A laser beam enters through a window in the top of the can, heating up the deuterium-tritium fuel to 3 million degrees Celsius (5.4 million degrees Fahrenheit)—not hot enough for fusion, but hot. Then comes the pinch: eighteen mega amperes of current travel through the beryllium can, generating an outer invisible can made up of mag-

*Frogs aren't usually magnetic, but with a strong enough magnetic field, they can be. Physicist Andre Geim won an Ig Nobel Prize (a satiric prize given to honor unusual, imaginative, even trivial achievements in scientific research) for demonstrating this in 2000; ten years later he'd win the real Nobel Prize for Physics for work on graphene, a sheet of carbon that's just a single layer of atoms thick.

netic fields. The current and the fields interact, and the pinch comes, squashing the can and the fuel with it for more than one hundred nanoseconds. The squishing causes the magnetic field inside the can to be amplified, reaching an extraordinary ten thousand Tesla. And, as is intended, the density and temperature ramp up too, with the nuclei reaching 30 million degrees Celsius (54 million degrees Fahrenheit). Fusion occurs (neutrons have been detected), and the fast particles that come screeching out of the reaction are bound by the magnetic fields so that they stay in the fuel, heating it further.[24]

It's a great idea for a fusion scheme. Sandia's simulations from 2012 show that a current of sixty mega amps would give a fusion gain of one hundred times the energy put in. Not so fast though! This is plasma physics after all. While Sandia doesn't have a machine big enough to try this out, even the preliminary experiments on Z showed that can compression was rife with instabilities. Just as with every other fusion scheme under the Sun, the practice is harder than the theory.

There are no fusion schemes without enormously complex physics and real head-scratchers to overcome: there are just fusion schemes that haven't been fully explored yet. Magnetized inertial fusion is promising, and a good example of how even supposedly slow-moving government laboratories can innovate, but it's got a long way to go to work out the kinks.

It doesn't have to be competitive. Because they have different strengths and weaknesses, the best solutions are likely to come from the private and public fusion sectors working together. After all, SpaceX has flourished because of its relationship with NASA, not in spite of it. Both types of star builder need to work together to create a regulatory environment that recognizes that fusion is distinct from fission (and the US Congress is listening).[25]

Ian Chapman is, sensibly, convinced of the need for public-private partnership.

"Five years ago we went to Rolls-Royce and firms like that and said 'can we work with you.'" At that time, the firms said no because they thought fusion was too early stage, he explained. "Now they're coming to us and asking how they can win fusion contracts," he continued. "You need the private sector for investment. A billion in the last ten years raised from private companies; venture capitalists, philanthropists, sovereign wealth funds, and, recently, energy companies: oil and gas firms."

Currently, the new star builders are behind the government laboratories in reaching a 100 percent energy gain from fusion. None have come close to JET's fusion energy record, nor NIF's best single-shot energy yield of 3 percent. Equally, they haven't had decades to do fusion, or the same level of funding. They're making rapid progress, though. And, if nothing else, they're forcing everyone to up their game.

So the race to build a star may yet be won by a maverick working in an unassuming warehouse somewhere near you.

And if you think that someone tinkering with a nuclear device in your backyard sounds, well, just a little bit *dangerous*, that's understandable. When I asked her what people should know about the race to build a star, Sibylle Günter shot back, "It's difficult— don't try it in your backyard." She does have a point. After all, the pistol shrimp's claw is a weapon. So let's find out if we need to worry about what our nuclear neighbors, the star builders, may be up to in your backyard.

CHAPTER 8

ISN'T THIS ALL A BIT DANGEROUS?

"The fact that no limits exist to the destructiveness of this weapon makes its very existence and the knowledge of its construction a danger to humanity as a whole. It is necessarily an evil thing considered in any light."

—Enrico Fermi and Isidor Rabi discussing the
hydrogen bomb in a report on behalf of the
General Advisory Committee of the US
Atomic Energy Commission, 1949[1]

One day before dawn in the early spring of 1953, Matashichi Oishi stood on the deck of the *Lucky Dragon No. 5* watching the fresh salty spray of the Pacific Ocean crash over the sides of the boat. He was twenty years old and working as a fisherman. It was the only life Matashichi Oishi really knew—he'd been doing it since he dropped out of school at age fourteen, and he had no other experience of the world. He and twenty-two other Japanese crew had been at sea for five weeks, sailing through nothing but open, empty ocean for most of that time.

The *Lucky Dragon No. 5* was a thirty-meter long (about ninety-eight feet) fishing boat that was barely seaworthy. There was little machinery, and everything had to be done by hand. It was hard, perilous labor. They'd been unlucky throughout the tuna-catching trip, beginning when they temporarily ran aground and continuing when they lost half of their trawling nets to the waters.

On the morning of March 1, 1953, they were attempting to make one last haul before returning home. As Matashichi stood on the deck with the sea breeze in his hair, and the milky radiance of billions of stars overhead, there was a sudden and all-pervasive bright light.

"It emerged from the horizon from the west, from the east, and from every direction," Matashichi Oishi said. "It wasn't a flashing, it was like the light was flowing and covering the whole area. I thought it was some kind of natural disaster."

The crew all stopped to look in awe as the night took its leave early. No one spoke as they watched the spectacle. Seven minutes later, a sound arrived: a deep rumbling, like continents colliding, that rose from the bottom of the ocean. The crew unfroze, and everyone, including Oishi, dived to the deck, some crawling to the cabin to hide. The sound passed, and it then went very quiet. But their ordeal was only just beginning.

Next came a biblical rain of stark white powder from the sky. It fell everywhere; onto the sea, onto the deck, onto the crew of the *Lucky Dragon*, and even onto Oishi's face.

"I didn't feel any danger," Oishi said, "because it didn't leave any mark."

It wasn't snow. Curious as to what it might be, Oishi tasted it. He realized it was ash. It was crunchy. What the unlucky inhabitants of the *Lucky Dragon No. 5* didn't realize yet was that they'd been showered in tiny flakes of radioactive coral. With the boat covered in white radioactive ash, the crew hauled up the fishing nets and set course for Japan and their home port of Yaizu. On the way back, they began to feel strange.

"We were dizzy and some people started having diarrhea. The places where white ash had landed started swelling and lots of large blisters began appearing."

The journey back across the Pacific, past the edge of the Marshall Islands where they'd last laid anchor, took several days. After four days, their hair began falling out.

As soon as the boat returned to harbor, scientists and physicians realized what had happened to the crew. They had acute radiation syndrome. They were all hospitalized. One crew member died within six months, the others spent a year in the hospital recovering, but they were all plagued by health problems long after. Oishi would get liver cancer. Many would have their lives shortened.

Around eighty miles west of where the *Lucky Dragon No. 5* had been sailing that day in 1953 was Bikini Atoll, a slight ring of coral protruding out of the ocean that was at the northern end of the Marshall Islands. At 6:45 a.m., the United States had exploded a fission- and fusion-powered hydrogen bomb called Castle Bravo that blew a one-mile-diameter crater through the atoll's coral ring. The device used an initial chemical explosion to drive together fissionable isotopes (special types of plutonium or uranium) so that they reached a critical mass at which a fission explosion occurred. The energy from the first fission stage drove a small sphere of deuterium and tritium to a high enough temperature and density for fusion reactions to start. This second, fusion stage triggered yet more fusion reactions in a rod of material and unleashed vast numbers of neutrons that caused even more fission reactions. Each stage released truly fearsome quantities of energy that combined to make it the largest bomb in human history up to that point in time.

Moments after the explosion, the fireball had consumed everything within miles. Coral from the atoll that was blown into the air was heavily contaminated by neutrons, making it radioactive. Surrounding islands were showered with radioactive debris, and their residents were, eventually, evacuated. A one-hundred-mile-long plume of fallout spread eastward from Bikini Atoll to Matashichi Oishi and the other fishermen.

There'd been an exclusion zone around Bikini Atoll, but, tragically, it hadn't been big enough. Kenneth Ford, one of the original designers of the hydrogen bomb, was on the team that put together Castle Bravo. It had been his job to do the calculations that would predict how much yield, in energy, the bomb would release. Working with primitive computing and the knowledge of nuclear reactions then available, he came up with a figure of seven megatons; that's an explosive force equivalent to 7 million tons of TNT. The actual yield of Castle Bravo was fifteen megatons. That's a dangerous miscalculation, and it was the difference between life and death for some of the fishermen. Fifteen megatons is equivalent to one thousand of the atomic, or fission, bomb that was dropped on Hiroshima going off *at once*, in the *same place*.

The incorrect reckoning happened because some of the lithium-6 (lithium with six particles in its nucleus) in the assembly was actually lithium-7, which Ford expected to be inert, but which can capture neutrons and turn into tritium in the right conditions. Tritium can undergo fusion and release energy. As it does so, it produces neutrons that can cause more fission reactions, releasing more energy, and so on. The result was a much bigger explosion than anticipated, and the terrible events that befell the crew of the *Lucky Dragon No. 5*. The US was eventually forced to pay compensation.[2]

"It does not seem possible, at least in the near future," the great nuclear physicist Enrico Fermi wrote in 1923, "to find a way to release these dreadful amounts of energy—which is all to the good because the first effect of an explosion of such a dreadful amount of energy would be to smash into smithereens the physicist who had the misfortune to find a way to do it."[3]

Fermi was wrong (unusually) and we've had the ability to smash ourselves into smithereens using nuclear reactions for decades. Clearly, nuclear technologies are a double-edged sword that can be used to help or harm humanity. We've heard a lot in this book about how they might be able to help. Now we're going to ask whether nuclear fusion can be harmful. The sharp side of nuclear technologies is so dangerous that the word "nuclear" on anything can put people off. (That's why even life-saving nuclear magnetic resonance imaging machines lost the word "nuclear" as they began to appear in hospitals.) Star builders want us to embrace their favored nuclear technology—fusion for energy—but we need to know the risks. And there's no getting away from the fact that fusion is a key part of the most devastating nuclear technology in existence.

Nuclear weapons have changed war more profoundly, and more suddenly, than anything in history. Unlike the awful loss of life and property in the 1945 firebombings of Tokyo (100,000 dead) and Dresden (25,000 dead), total annihilation can now be achieved with just one bomb. The atomic, or fission, bombs used on Hiroshima and Nagasaki created a Hell on Earth. At Hiroshima, all living creatures and plants within half a mile of the blast center were immediately vaporized, leaving only black charred residue. Fourth-degree burns that penetrated the skin occurred even at a distance of 2.5 miles from ground zero. Eyewitnesses reported that victims' skin hung off their bodies. More than 90 percent of the buildings in Hiroshima were gone. Those who didn't die from the blast were engulfed by the subsequent fires, and many who survived both were killed by radiation sickness. The total death toll was probably 200,000. The bomb dropped on Nagasaki probably killed another 150,000 people. The destruction wrought by nuclear weapons is so effective at making entire cities vanish that, paraphrasing Arthur C. Clarke, it's as if the technology were magic, albeit of the darkest and most terrible kind.

Nuclear weapons are among the most dangerous threats to humans. A big enough nuclear war would likely plunge the world into a nuclear winter in which the Sun's rays would be blocked by soot and debris, temperatures would dip to levels not seen for twenty thousand years, and those not killed immediately would likely starve to death. There are more than ten thousand nuclear weapons in existence today. We must all do what we can to ensure that they're never used.[4]

There are two types of nuclear weapons, those based on fission (atomic bombs) and those based on fusion and fission (hydrogen bombs). Hydrogen bombs, like the one that obliterated Bikini Atoll in 1953, are *much* more deadly than atomic bombs. The arsenals of the world's nuclear powers are now stocked with them.

But it would be incorrect to assume that peaceful nuclear fusion designed to provide energy poses the same kind of danger.

Happily, a fusion reactor just can't be rigged up to explode like a hydrogen bomb. Controlled fusion reactors designed to produce power are fundamentally different from hydrogen bombs. The proliferation of nuclear weapons always requires fissile material: you can't build either kind of nuclear weapon without the isotopes that undergo nuclear fission, usually special types of uranium or plutonium. In an atomic bomb, fission of these isotopes provides the explosive energy. In a hydrogen bomb, fission provides the initial trigger that kicks off the fusion reactions. Fusion reactors, however, involve no fission at all—so they don't work like a bomb of any kind, and can't be made to.

A more legitimate concern is that a fusion reactor could be used to create the material to assemble a nuclear weapon. It's scientifically possible for someone to take uranium-238 or thorium-232 into a fusion power plant and expose it to fusion-produced neutrons to breed enough fissionable plutonium-239 and uranium-233, respectively, to get the basic materials for a fission bomb. One study calculates that the fastest this could be done with no attempt to hide what was going on and with full access to the uranium and thorium precursors would be a period of months. I asked the star builders whether this meant fusion energy was a risk for nuclear proliferation.[5]

"It's always a risk," Dr. Mark Herrmann, NIF's director, conceded, "but the advantage with fusion is there's no need for any fissile material to be around." His strong belief is that fusion is safer from a proliferation point of view than fission, and it's a view shared by everyone I spoke to at Lawrence Livermore. Don't forget that one of the laboratory's primary missions is preventing the proliferation of nuclear weapons. The other star builders felt the same: "Fusion is far, far lower risk [than fission]," CEO of Tokamak Energy Jonathan Carling said, "because it doesn't involve any fissile material like uranium or plutonium."

There's no need to have any materials at a fusion reactor site that could be used either to build an atomic bomb or as the trigger for a hydrogen bomb. "You can't break into a fusion power plant and steal some stuff and build a bomb with it," Dr. Nick Hawker, CEO of First Light Fusion, told me. Besides, fissionable isotopes and their precursors are detectable in small quantities, so it's relatively easy to check for them. Inspection equipment at a reactor would need only seconds or minutes to find out what was going on. And all of the materials needed to breed the isotopes for a bomb are—as Nick Hawker pointed out to me—very tightly controlled.

So while the star builders agree that there's a risk of proliferation, they believe that it's far lower for fusion than for fission. In the case of a fission plant, it could take weeks to openly and aggressively create the critical mass for a bomb, and more than one nation has been accused of clandestinely creating the ingredients for nuclear weapons using fission power as a cover. Although everyone agrees it is far harder to use fusion power as cover for a weapons program, star builders will need to bear the risk of proliferation in mind if they're serious about making fusion one of the planet's main energy sources.[6]

When it comes to all things nuclear, though, the concerns extend beyond proliferation.

"If you think about what worries people most," Jonathan Carling told me, "it's a meltdown, because when we've seen other events, like Fukushima or Chernobyl, it hasn't been good. Fusion plants can't have a meltdown." All star builders agree. "There's no risk," said Nick Hawker, "that a fusion reactor can melt down: if you interrupt it, it just stops."

Ian Chapman puts it like this: "In fusion, if you want to stop a reaction, it takes milliseconds; the hard thing is to keep it going. It's easy to stop, it's really easy to stop, so there is just no risk of chain reaction in the same way."

Meltdowns occur because, once they begin, fission reactions can be hard to stop. Fission reactions happen in a chain, with each reaction giving rise to the next. Enough uranium fuel is kept within the reactor chamber to keep this chain going. A fission power plant is balanced to keep the chain in check so that the number of reactions is sustained without growing explosively, or spluttering out and stopping altogether. Using the analogy of a pandemic, it's like keeping the effective reproduction number, R, at 1. In a fission power plant meltdown, the chain gets out of control, the heat rises substantially, and parts of the plant can literally melt.

You might worry that a fusion reactor could get out of control too, burning through more of its fuel than star builders intend. Yet because controlled fusion relies on temperature, confinement, and density, rather than a chain, fusion can't race out of control in the same way. In contrast to fission, fusion often needs energy to be put in to keep the plasma confined. In the case of magnetic confinement fusion, what activates the process is externally applied heating or magnetic fields. Stop them suddenly and you'd get a reactor disruption, but the plant wouldn't explode or melt down: fusion reactions are hard to start and easy to stop. In inertial confinement fusion, stop the driver—in NIF's case, that's the laser—and the whole plant shuts down. There's no runaway effect in play.

And, unlike with the fission process, there's only a tiny amount of fusion fuel in a fusion reactor at any one time, so the maximum amount of energy that can be released is severely limited. The biggest planned tokamak will have, at most, two grams of hydrogen in it at any time; far less than is present in a nuclear weapon. Putting any more hydrogen into either a magnetic or inertial fusion machine would stop the conditions for fusion from being reached: the reactors are naturally self-limiting. Tokamaks can only run with a certain hydrogen density—any more and they can't get hot. And, in inertial confinement fusion, there's just not the driver energy to implode capsules much bigger than the ones that are currently being fielded. The only type of fusion in which the whole fuel supply is kept within the reactor is gravitational confinement fusion in stars.[7]

The star builders are keen to stress that fusion's safety is part of why they're so passionate about it. Fusion reactors can't be turned into bombs, undergo uncontrollable chain reactions, or melt down, because they don't involve nuclear fission. What fission and fusion *do* have in common, though, is radioactivity, which poses a genuine danger.

Radioactive!

Radioactivity is perhaps the most misunderstood aspect of nuclear technology, and it's a big reason why many people are suspicious of nuclear power—especially nuclear fission.

Radioactivity seems mysterious and dangerous partly because it's invisible to the naked eye. Although fire or flooding are much more common risks to human life, they at least can be seen, understood, and avoided. To understand the risks of radioactivity from fusion, let's start with what radiation actually is.

There's particle radiation, formed of high-energy electrons, protons, neutrons, and other fundamental particles. And there's light radiation, such as X-rays or gamma rays. Activated, or radioactive, substances have atoms that are unstable and will occasionally decay into other atoms. Physicists often think about radioactive materials in terms of the time it takes for half of the atoms to decay, the half-life. Fissile uranium-235, for example, has a half-life of around 700 million years.

When radioactive atoms do decay, they often ping out one of these types of radiation. The weak nuclear force is behind these decays. The radiation that is described in textbooks is usually said to be one of three types: fast helium nuclei (alpha radiation), electrons (beta radiation), or very energetic light (gamma radiation). Alpha, beta, and gamma. But, guess what, put a load of energy into any particle and it can still cause lots of damage—a bullet is only harmful because of how fast it's going.

So I don't think the alpha, beta, gamma names given to radiation are particularly useful, and they mostly exist for historical reasons because those types of radiation were discovered first, and they're the most common.[8]

A better way to think about radiation is that it's composed of particles or sometimes packets of light that have enough energy to break up atoms and molecules as they go, potentially creating more radiation in the process. For instance, the carbon-14 found in tooth enamel decays when a neutron in its nucleus becomes a proton, pinging out an electron (beta radiation) as it does. The fast neutrons created by nuclear fusion reactions can also change atoms and induce radioactivity.

It may come as a surprise to many people, but radiation is completely natural. Low doses pose almost no danger and are unavoidable. You're being strafed by radiation as you read this; approximately every single second of every day, every square meter of the planet is hit by hundreds of high-energy "cosmic rays," bits of radiation that originated *outside* our solar system. Scientists are still not 100 percent sure how cosmic rays get all the way to Earth at such high energies, but they're a normal part of our universe. When they hit the upper atmosphere, they can create cascades of radioactive isotopes, including carbon-14.[9]

Radioactive substances can also be found in the earth, and the early radioactivity pioneers, such as Marie Curie and Ernest Rutherford, did their first experiments on these. These ores originally come from outer space and were created in supernovae. You may wonder how, if these atoms are unstable, and formed in supernovae, they haven't all decayed yet. The instability of some of these "unstable" atoms is not instability as we would understand it in our everyday lives; the atoms decay over a period so long that it is hard to imagine—as long as 20 billion billion years.[10]

The extent to which radiation released by radioactive material is dangerous for biological creatures like ourselves is measured in "dose equivalent," the amount of energy dumped into your body as radiation goes through it. It's counted in sieverts, or Sv. What is or isn't radioactive, and how radioactive a given thing is, is quite surprising. A banana, for example, doses you with one-ten-millionth of an Sv because of the potassium it contains. A single chest X-ray is roughly twenty-millionths of an Sv. An eight-hour flight from London to New York is forty-millionths of an Sv.

Most people receive around 4 milli Sv (mSv) a year—one hundred times more than a single flight from London to New York— just from natural background radioactivity and medical scans.[11] ("Milli" means thousandth, as in millimeter.) Cosmic rays are responsible for 10 percent of annual exposure. At that dose, radiation is not dangerous. To pose a risk to health, the doses must be either much larger or be received over a short period. It might be a surprise, but in normal times and with safe operation, even the radioactivity from being near a nuclear fission plant is negligible. For example, in the UK, nuclear fission workers' annual occupation-related exposure is just 0.18 mSv.[12]

A dose of 100 mSv per year or more is associated with an increased risk of cancer, and symptoms of radiation sickness would usually only appear after doses of a few hundred mSv over a short period of time. It's useful to put these numbers into some perspective. Two of the most famous nuclear fission disasters are the Three Mile Island incident and the reactor meltdown at Fukushima. In the former, some people were exposed to 1 mSv, and those in the Fukushima exclusion zone in the two weeks after that meltdown would have had a 2 mSv dose—far lower than the level associated with cancer. However, six workers at the Fukushima plant received higher doses than that level, and tragically, there has subsequently been one death that can be directly linked to radiation. The doses pale in comparison to those received at Chernobyl, the world's very worst nuclear meltdown, however. Thirty people received fatal doses of more than 8 Sv—that's eighty times the cancer limit and eight thousand times the dose at Three Mile Island. But it's not just nuclear power that is associated with radioactivity. Coal power plants release uranium and thorium into the atmosphere as they burn fuel. Some analysis estimates that coal ash carries one hundred times more radiation into the environment than a safely operated nuclear power plant producing the same amount of energy.[13] So radiation is everywhere, and the doses can vary widely both in their magnitude and in the extent of harm they cause.

This all sounds very depressing. But radiation isn't all bad, far from it. It can be really useful. It's commonly used to sterilize food. There's a good chance that it's keeping you safe right now. Within most smoke detectors is a small amount of radioactive matter that emits helium nuclei. Composed of two protons and two neutrons, helium nuclei are positively charged, and cause other particles in the air to become charged when they crash into them. The stream of helium nuclei in your smoke alarm sets up a small current that can be detected. When smoke is present in the air, the helium nuclei get absorbed by it, less current flows, and the alarm goes off.

Radioactive carbon-14 is fantastic for accurately dating organic objects that are thousands of years old. When animals or plants die, the amount of radioactive carbon-14 in their bodies is fixed at a known fraction. Just as with teeth, you can work backward to find out how long ago it must have been that the plant or animal died. This technique works for up to around fifty thousand years, after which there's too little carbon-based radioactivity to measure. Evidence from radiometric dating has revolutionized our understanding of human prehistory in all kinds of ways. Thanks to very long-lived radioactive sources, Rutherford was able to argue in a 1904 lecture that the Earth was billions of years old (geologists already realized this, but the physicists hadn't been so sure). The latest radiometric dating uses the decay of uranium to lead within zircon crystals to show that the Earth is at least 4.38 billion years old.[14]

And radioactivity can be used in medicine to diagnose or treat various diseases. One of the pioneers of this technique, winning a Nobel Prize for it, was Ernest Lawrence, after whom Lawrence Livermore National Laboratory is named. The technology he developed helped to slash the price of radioactive isotopes used in cancer treatment from $100,000 a gram in 1921 to just a few dollars a gram in 1935, and that was just for the salt that had to be put into the machine.[15]

Another enormously useful nuclear diagnostic involves a patient purposely ingesting radioactive isotopes. Once inside the body, the absorbed isotopes reveal their location by emitting X-ray or gamma radiation. Like many technologies, radioactivity can be a tool or a terror.

It's the terror that more often makes the news. In the case of the *Lucky Dragon No. 5*, which, unluckily, found itself too close to the Bikini Atoll blast site, it wasn't the explosion itself that did the damage, it was the radioactive rain. Fortunately, large scale radiation exposures are very rare. Meltdowns don't happen very often.

What many more people are rightly concerned about when it comes to radioactivity and nuclear power is what goes into and, even more so, what comes out of the plant.

In fission, both the inputs to and outputs from the reactor are radioactive. Let's take the "ingredients" first. The most potently fissile commonly occurring isotope is uranium-235, which is also radioactive and decays into thorium by emitting a helium nucleus. Importantly, its long half-life means not very many atoms decay per second, so it's not the most dangerous radioactive material by a long shot.

What comes *out* of nuclear fission reactors is more problematic: radioactive waste that can last for millions of years. In fission, the splitting up of large nuclei like uranium-235 creates gamma rays, an energetic neutron, and a bunch of smaller atoms that are themselves likely to be unstable. French nuclear company EDF Energy estimates that the yearly amount of nuclear waste per person in France (which generates 75 percent of its electricity from fission) is one kilogram (approximately thirty-five ounces). It's a lot less waste by mass than fossil fuels produce, but what *is* created is more problematic. Most of the one kilogram is low-level waste that becomes safe relatively quickly. The really dangerous part is the 3 percent of high-level waste so radioactive that it must be actively cooled for its first forty years. It then needs to be carefully secured and stored for at least a thousand years. Only after one hundred thousand years does this waste reach the same radioactivity level as uranium ore, which can be handled with gloves.[16] Countries around the world are still grappling with where to store this long-lived waste. For now, it's mostly stored next to the plants where it was created.[17]

It's understandable that people would be concerned about the small though dangerous and long-lived waste generated by nuclear power. But, star builders contend, nuclear *fusion* is different from *fission*, even when it comes to radioactivity. Even if we assume fusion-produced waste is equally as dangerous as waste from fission (we'll come to whether it is shortly), the most serious incidents in which people were exposed to radioactivity from fission power have all happened because of meltdowns—occurrences that can't happen with fusion power plants.

The inputs to fusion are arguably safer too. Deuterium isn't radioactive. However, tritium, the other ingredient of fusion reactions, is. It has a half-life of about twelve years, and decays into helium-3. But tritium is so weakly radioactive that the electron it emits when it undergoes this decay can't even pass through the dead layer of skin that surrounds your actual skin. Because of this, it has found some curious uses, including in a radioactivity-powered light that requires no batteries. Another use is for illuminating emergency exit signs; even when the power fails in a building, the tritium will provide a radioactive glow. Tritium is fairly safe as radioactive substances go, but it needs to be handled with care.

I asked the star builders whether the stocks of tritium kept on-site at a fusion reactor could pose a risk to anyone. Lorne Horton, who works at the Culham Centre for Fusion Energy, says that the radiation risk from fusion fuel at JET is low: "We work at industry levels, which are smaller doses than you get from just living every day and certainly smaller than what you get from taking an airplane ride. My dose is unmeasurably small."

I put it to Ian Chapman that residents of Culham might be concerned that some of JET's tritium could escape.

"In terms of the radiological exposure in the event of a big accident and loss of containment, we're looking at a few tens of grams of tritium," he said. "It's a negligible risk to the population: we have about one hundred grams of tritium here on site and we're not under a nuclear license, we operate under the Environment Agency. It's inside a containment vessel, inside another containment vessel, inside a concrete bio-shield—but even if all three were penetrated then it would go into an exhaust de-tritiating system that would catch it. If that were to also melt at the same time and it went out into the atmosphere, the risk is negligible because the inventory is so low."

Because of these low risks, fusion plants, unlike fission plants, have no exclusion zone around them. The fusion start-ups' levels of safety procedures vary depending on whether they're at the stage of doing deuterium-tritium fusion or not, and for now, most aren't and so have no need for tritium (yet). Even for the biggest planned tokamak in the world, which is likely to be on the same scale as a working fusion power plant, there is no scenario that would necessitate evacuating locals—and the radiation from a plant that's running will be a thousand times less than the natural background level of radiation.[18]

"When it comes to working with ionizing radiation, the regulations are very proscriptive and we follow all of them," Nick Hawker assures me. "The total inventory of radioactive material we'll have for the gain experiment will be tiny in tritium; there'll be zero of anything basically. For the power plant, there'll be tritium, and it's the biggest worry as it's mildly radioactive and it's mobile." He went on to explain that the safety requirements for tritium are much less onerous than the ones for fission fuel.

The fusion of deuterium and tritium directly produces fast-moving neutrons and helium nuclei, albeit in very small quantities. The helium nuclei from fusion quickly recombine with electrons as soon as the power is removed from a fusion reactor, so, in that circumstance, they pose no radioactivity problem. In fact, helium is useful, and in short supply.

The biggest source of radioactivity from fusion reactions is the neutrons that are generated when the reactor is on and working. It is for this reason that everyone working near the reactor cores at NIF and JET has to wear dosimeters that check how much radiation they're being exposed to.

As Dr. Bruno Van Wonterghem, the operations manager at NIF, took me around the facility, he pointed out some of the precautions they're taking to stop neutrons from harming anyone. NIF's aluminum target chamber, where the laser meets the fusion fuel, is itself encased in a concrete cylinder six feet thick accessed by four-foot-thick doors. *This* cylinder is encased in another concrete wall, and there's further shielding after that too. During a shot, most of the ten thousand million million neutrons created by fusion reactions escape the target chamber. As with JET, the concrete layers that surround it are packed with boron, an element that safely absorbs neutrons. Each wall of boron-rich concrete reduces the number of neutrons escaping by a factor of one thousand.

"Neutron doses in the target bay [the space immediately outside the steel target chamber] during high-yield shots are about eighty thousand rad," Bruno told me, calmly. "They're beyond lethal." The rad is a measure of how much radiation your body absorbs. Sieverts also take into account that even for the same energy, some types of radiation are more damaging. For the fast neutrons produced by fusion, eighty thousand rad is equivalent to 800 Sv. This is an extraordinary amount as just 5 Sv will kill you. There's a reason there's enough shielding in place in just the first twenty meters (approximately sixty-five feet) from the reactor chamber to reduce the number of neutrons a million times over, with the dose falling even more. The multiple stages of shielding that enclose the target chamber reduce the dose to levels that pose no risk to anyone, as my radiation badge attested when I witnessed a shot from the nearby control room. And Bruno told me that the target chamber has enough shielding to cope with a fusion gain of sixty—that is, sixty times as much energy out as went in.

We walked through one of the thick steel-frame doors, also filled with concrete, and switched from the non-radioactive zone to the radioactive zone. We were entering the target bay. In the center was the reactor chamber, inside which the laser beams meet the hohlraum. While the most lethal doses occur during a shot, radioactivity induced by the neutrons can linger. I was utterly amazed that we were able to wander around immediately outside the target chamber when the area has been exposed to beyond-lethal radiation so recently. Bruno did remind me not to touch anything. Wearing the appropriate attire, engineers were moving pieces of recently irradiated equipment around without fear. When I mentioned this, Bruno told me that it took anything from a few hours to a dozen hours after each shot for the radioactivity to fall to the background level everyone experiences every day.

I asked the star builders how much of a risk the radiation from a typical fusion experiment poses.

"Not much," Mark Herrmann told me, speaking about NIF. "The chamber is very big. The doors are a couple of meters thick for high-yield experiments. There are two sets of doors. You can sit in the control room even on the highest-yield shot."

The reason Mark can confidently say this is because we know how to stop radiation if we need to. Even though a fusion-produced helium nucleus travels some twenty-five thousand times faster than a speeding bullet, it can be stopped by a sheet of paper. A few millimeters of aluminium will stop fast-moving electrons, while gamma rays or neutrons will only be stopped by a big block of a dense material, like lead. The biggest problem by far for fusion is the neutrons—and they're why there are multiple layers of concrete at NIF and JET.

However, it's not just humans that are affected by radiation. Over a long time, the repeated bombardment of reactor chambers by neutrons has consequences. When fast-moving neutrons crash into atomic nuclei, they can cause them to become radioactive. The type of atom that is struck is what determines whether this will happen or not. Star builders have to choose the construction materials of their reactors carefully so as to minimize neutron-initiated activation. Even with the most resilient materials, over years of being strafed by neutrons, some material will become activated. This means that fusion produces radioactive waste—but it's very different from the waste left by nuclear fission reactors.[19]

"Fission plants produce waste as an inherent part of their process," Jonathan Carling told me. "The only waste that a fusion plant produces is relatively low-activation waste of the plant itself because it gets hit by neutrons." In fission, the spent fuel and the reactor are radioactive; in fusion, it's only the reactor chamber that becomes radioactive, and that will be dealt with at the end of the plant's life.

So how easy would it be to clean up a nuclear fusion reactor site at the end of its life? JET's been running since 1983, including with neutron-producing deuterium and tritium. It's the closest we've come to a working fusion reactor and it serves as a good model of what to expect.

"When we close JET, the plan is to be greenfield within ten years," Ian Chapman told me. "Robots go in and remove the first walls, walls get de-tritiated . . . we store the tritium, throw the rest in a skip [a dumpster]. The residue is negligible. To say we've been operating for thirty-five to forty years and go to greenfield in ten years gives you a sense."

How much radioactive waste fusion reactors will create at the end of their lives depends on what the reactor chamber is made from and how long it has been in operation, but the best estimates suggest that after ten years, a nuclear fusion reactor chamber that has been creating ten times as much energy out as put in would be no more radioactive than uranium ore. After one hundred years, it wouldn't be radioactive at all.[20]

"You can put it into intermediate waste dewars [sealed flasks] until it decays and then get rid of it," Ian Chapman says. "You don't have the same legacy that you do with fission."

Radioactivity is good and bad, common and rare, dangerous and safe: it all depends on the context. Star builders have good reasons to believe that fusion reactors will pose fewer radiation risks than fission plants and leave no lasting legacy of radioactive waste.

The Real Dangers of Fusion Reactors

If it's not radiation, nuclear proliferation, or meltdown, what *is* a big problem for star building?

"The biggest risk," Lorne Horton, JET's exploitation manager, told me, "is that you fall down." He meant this literally. Then there are the risks of being near large electricity supplies and powerful magnets, he continued. In his reckoning, radiation comes a distant third. Without any prompting, Nick Hawker also told me that working at heights was First Light Fusion's biggest safety concern.

"We're not dangerous to the general public," he explained, "and with proper process, which we have, it's not dangerous to the team either."

The truth is that we won't know just how safe a working, power-generating nuclear fusion reactor will be until the star builders get around to building one. We know that a working fusion reactor is likely to be somewhat safer than a nuclear fission plant, because fusion reactors can't melt down and they involve fewer risks from radioactivity. But the two technologies have some similarities too—they're both strictly regulated, use little fuel because nuclear energy is so efficient, and need a great deal of complex infrastructure in a small area.

So as an upper bound on just how dangerous nuclear fusion will be as a power source, I thought it would be interesting to consider its cousin, nuclear fission. And that has led to a finding that may come as a big surprise to many.

Nuclear fission power stations are the safest large-scale form of energy production on the planet. For every exajoule of energy generated by fission (and remember that the US alone consumes ninety-five exajoules of energy each year) there are around 20 deaths. To put that number into perspective, an exajoule of coal results in 6,800 deaths. By this simple metric, fission is 340 times safer than coal. It makes more sense than it might first appear; you need just two grams (less than an ounce) of uranium for every sixteen kilograms (approximately thirty-five pounds) of fossil fuels, which means a lot less mining and extraction of ores, one of the riskiest activities for fossil fuel power.

All power sources come with risk, even energy from renewable sources. Hydroelectricity is associated with 330 deaths per exajoule, a number dominated by a single, terrible event—the failure of the Banqiao Dam in China, in 1975. It killed an estimated 200,000 people. Likewise, the 20 deaths per exajoule for fission takes Chernobyl into account.[*] Putting aside uranium extraction and looking at 1990 onward, fission is associated with just 3 deaths per exajoule. For wind and solar, it's 10 and 5 deaths per exajoule, respectively, yet another reason why solar energy is a promising part of the solution to the energy crisis.[21]

Nuclear disasters are big news precisely *because* they're so rare. It's akin to air disasters and car crashes: each year, car crashes kill tens of thousands of people in the US. Worldwide, plane crashes typically kill fewer than two thousand. But which do we hear more about? This asymmetry can result in poor policy choices—after the meltdown at the Fukushima nuclear plant in Japan, Germany decided to phase out its entire fleet of nuclear fission power stations.

---

*But not Fukushima, because it happened after the study was published.

In consequence, Germany was forced to replace nuclear energy with burning fossil fuels. What happened? An estimated additional 4,600 deaths and three hundred megatons of carbon dioxide emissions between 2011 and 2017. Japan also shut down all of its own nuclear power plants in the immediate aftermath of the Fukushima accident. This caused a significant rise in energy prices, and people responded by buying less energy during winter. As a consequence, more people died due to energy poverty—in fact, the mortality from this effect was higher than from the accident itself. We should look carefully at the evidence before overhauling policies based on instinct alone.[22]

Fission can save lives if it displaces more dangerous power sources. Throughout its history of use, fission power has, it's estimated, prevented 1.8 million air pollution–related deaths and sixty-four gigatons of carbon dioxide–equivalent greenhouse gas emissions that would have resulted from fossil fuel burning.[23]

Again, all power generation involves risk; it's inescapable. As a society, every time we use energy, we make a trade-off; we decide that the risk of mortality or morbidity is worth it for the benefits the energy offers. Our primary power source today, coal, offers a bad trade-off. Despite a small number of prominent nuclear disasters and the risks related to radioactivity, nuclear fission offers one of the best trade-offs out there.

Nuclear fusion, star builders reason, would be substantially safer than fission because a fusion plant can't melt down and doesn't involve long-lived, high-level radioactive waste. Most important, fusion would generate large amounts of energy from even smaller amounts of fuel than are required for fission. If the star builders are right, then nuclear fusion, when it's ready, will be one of the safest—if not *the* safest—large-scale power sources on Earth. Maybe fusion is lucky after all.

CHAPTER 9

FINISHING THE RACE FOR FUSION

"The day when the scientist, no matter how devoted, may make significant progress alone and without material help is past . . . Instead of an attic with a few test tubes, bits of wire and odds and ends, the attack on the atomic nucleus has required the development and construction of great instruments on an engineering scale."

—*Ernest Lawrence, Nobel Prize banquet speech, 1940*[1]

"Fusion has always been thirty years away and that gets thrown at me at every public outreach talk I do," Culham's Professor Ian Chapman says when I ask him if star builders are closer to delivering net energy gain than they were thirty years ago. "It ignores the huge progress that's been made."

While Ian talks candidly about mistakes, he believes that we're closer than ever to building a star. The data back him up: between 1957 and 2018, the joint combination of temperature, density, and confinement achieved by any star machine increased by a factor of 1 million.

JET deserves credit for having pushed progress the furthest with its brief generation of 16 megawatts and a Q of 67 percent in

1997—tantalizingly close to the finish line. Most machines have only managed fractions of a percent.

Although JET's record hasn't been broken, progress on parts of the problem have continued apace. The very first devices controlled plasmas for microseconds. In the late 1950s, John Cockcroft's ZETA machine improved that one thousand times over. JET, built in 1984, runs for a few seconds—another thousandfold improvement. But controlling fusion plasmas for *minutes* is now possible. In 2003, the Tore Supra tokamak in France managed to control plasma for six minutes and thirty seconds. Also on the accomplishment list: many tokamaks are now able to reach temperatures beyond a hundred million degrees. The EAST tokamak, in China, has managed to combine high temperatures, as high as fifty million degrees, with times of one hundred seconds. Both EAST and France's WEST tokamak are aiming to control their plasmas for longer than fifteen minutes. Records keep being broken: in 2020, the Korea Superconducting Tokamak Advanced Research (KSTAR) managed twenty seconds at more than a hundred million degrees.[2]

The progress on which JET relies, and future tokamaks will build, has largely been won by ever bigger machines, ever bigger teams, and some serendipity. "The present status of fusion is primarily the outcome of a large effort and a lot of dedicated work, and only at a few occasions did we obtain results which were so pleasantly surprising that the community even took time to accept them," Professor Sibylle Günter, scientific director of the Max Planck Institute for Plasma Physics, told me.

Sibylle's list of the breakthroughs begins with the Russian T-3 tokamak that appeared on the international scene in 1969 and completely changed the direction of magnetic fusion research. She also talks about a discovery at her own institution in 1982. A researcher named Fritz Wagner working on the ASDEX tokamak in Garching found machine settings that almost magically caused both the confinement time and plasma density to *double*. Without this single discovery, tokamaks would probably have to be twice as big.[3]

Other progress has come from theory and computation. It's much easier and cheaper to do an experiment on a computer than in a $2.6 billion star machine, if the simulations are realistic. As both computer hardware and software have improved, so simulations have become better guides to reality. Researchers at JET recently used machine learning to predict disruptions tens of milliseconds before they occurred, giving them time to tweak the plasma to dissipate the buildup of energy and forces. Happily, the insights garnered can be applied to other tokamaks. Such breakthroughs are a step toward more reliable reactors.[4]

Ian Chapman thinks these improvements could give the decades-old JET a shot at setting a new world record for fusion energy when it begins a new run of deuterium-tritium experiments. "When we last ran JET with deuterium and tritium in 1997 it was a point in time," he says. "The power raced up and we lost control of the fuel, and it was a millisecond. We're going to do that again, but this time we'll race it up and hold it there, and bring it down in a controlled way . . . We now understand plasmas better than we did twenty years ago."[5]

Magnetic fusion machines are edging closer and closer to the finish line. If magnetic fusion star builders can get the combined temperature, density, and confinement times just another 1.8 times higher, they'll hit ignition. That's when their plasmas not only get net energy gain, but generate enough energy from fusion to remove the need for external heating.[6]

Far out in the lead in achieving the trio of temperature, density, and confinement simultaneously are three state-sponsored tokamaks: Japan's JT-60, South Korea's KSTAR, and the Joint European Torus (JET). The stellarator Wendelstein 7-X is not far behind them, achieving impressive feats for a relatively new machine based on a different technology.[7]

For now, it looks as though the private sector fusion firms are lagging behind—though some are keeping the details of their progress secret. At the back of the pack are those firms aiming to do fusion reactions that aren't deuterium-tritium: they have a very, very long way to go because the conditions required are more extreme. On the other hand, many private sector star builders have been playing an impressive game of catch-up. The real question for them is whether they'll be able to continue the rapid rate of progress and leapfrog ahead.[8]

The huge progress made by magnetic confinement fusion machines means that net energy gain is within reach. And even if JET can't make it all the way, another machine is coming that almost certainly will.

We're Going to Need a Bigger Tokamak

There's one aspect of the current fleet of magnetic fusion machines that is holding back progress. It's a lesson that has been learned time and time again in fusion, a lesson that the Big Bang, stars, supernovae, and nuclear weapons have been telling us all along: fusion works best on big scales. For conventional tokamaks, the confinement of plasma gets better the bigger the machine is. Going from JET, at a three-meter (approximately ten-foot) radius, to a tokamak twice that size would improve confinement four times over.[9]

In 2006, recognizing the need for a machine that can run with deuterium and tritium and has the necessary scale to go beyond net energy gain, a group of thirty-five countries got together and, in a ceremony at the Elysée Palace that was attended by four hundred guests, signed an agreement to build an almighty tokamak in the South of France, in Cadarache, near Marseille. The members of this thermonuclear club are China, the European Union, India, Japan, Korea, Russia, and the United States, with the brunt of the costs borne by the EU. These nations represent more than half the world's population. It took them twenty-two years of planning, arguing, design, and preparation to agree to build the new machine: ITER.

When it is completed, ITER will be the world's largest tokamak, and one of its key objectives will be to demonstrate net energy gain. It's a behemoth. Much of it is designed on principles that have been tested on JET, but it will operate with more than ten times the volume of plasma—and bigger plasma volume usually means more stability. Instead of having a three-meter-radius torus, ITER will have one that is more than six meters.

ITER's engineering is more complex than that of any other machine yet built. The tokamak's superconducting magnets will need to be formed from one hundred thousand kilometers (approximately sixty-two thousand miles) of niobium-tin wire. Each finished magnetic coil will stand seventeen meters (about fifty-six feet) tall and nine (about thirty feet) wide and be cooled to four degrees above absolute zero. Eighteen of these will be arrayed around the tokamak to generate a magnetic field of thirteen Tesla (3 million times the Earth's magnetic field), storing up tens of thousands of megajoules of magnetic energy. ITER will take up 180 hectares (equivalent to 250 soccer fields), and when finished, its structure will have a mass equivalent to three Eiffel Towers.

While ITER is situated in the South of France, the work being undertaken for it is truly international. There are more than fifteen satellite sites around the world providing expertise and parts. The project's current director general, Dr. Bernard Bigot, has described it as an "extraordinary human adventure" and he's not just talking about the three thousand or so staff who will be needed to run ITER; this is a fusion experiment on behalf of the planet.[10]

As with many large and complex international collaborations

where juicy contracts are available, construction has been beset by delays. The machine was initially due to turn on in 2016, but that target date has slipped by, and the first plasma isn't scheduled to appear until 2025, with deuterium-tritium experiments expected to begin even later, in 2035. As of the start of 2020, construction was two-thirds complete.[11]

ITER has been designed to win the race to build a star. It's aiming to hit a plasma Q of five—that is, five times as much power out as in. It will also try to reach a Q of ten for a few hundred seconds, generating five hundred megawatts of power for the fifty megawatts of power put in. The plasma will reach temperatures of 150 million degrees Celsius (270 million degrees Fahrenheit). The gain in energy won't be enough for a commercial reactor, and ITER won't deliver a single watt of power to the grid, but it's enough to show that fusion power is possible. If it works, ITER will shift fusion from fantasy to tangible power source.[12]

We're Going to Need a Bigger Laser

Inertial confinement fusion has also seen huge improvements, with NIF by far the leading machine. Many had hoped NIF would reach ignition when it was completed in 2010. The laser performed well. The experiments on nuclear stockpile stewardship that dominate NIF's laser shots are said to have been a success. Scientific experiments probing conditions in planets and supernovae have been a hit. But on the inertial fusion energy program, progress was initially much less forthcoming. Through 2011 and 2012, the implosions were riddled with instabilities. While neutrons were generated and energy released, the results were orders of magnitude less than what had been expected from parsimonious mathematical models and simple computer simulations.[13]

Following sharp criticism from Congress and the US Department of Energy, NIF's scientists changed direction and started trying to understand what was going wrong in the implosions, and why they weren't behaving like the simulations suggested. New staff—including new director Dr. Mark Herrmann and new chief scientist Dr. Omar Hurricane—were brought in and the program was refreshed. With new ideas, NIF found its stride. The experiments are now breaking records for inertial fusion, and they're not a million miles behind JET's world-best record for energy yield.

Between 2011 and 2019, the fusion energy released in the best shot on NIF rose sixty times over. In 2018, NIF first reached an energy yield of 3 percent. In 2020, NIF's scientists announced that they had exceeded this. This is too much fusion to be occurring just due to the energy coming from the laser. Making the analogy with fire, it's robust evidence that heat isn't just coming from the match. After a rocky few years, NIF is back in the race.[14]

Just as with the star builders using magnetic confinement, computers have played an important role in NIF's success. In 2018, Livermore's supercomputer, Sierra, became the second fastest in the world. Better computing means that it's possible to simulate more of the complex ways that plasmas can misbehave, or use machine learning to guide experiments.[15]

But the most significant progress has come from better control of how the laser and the plasma interact in the hohlraum, the gold box that houses the capsule of fusion fuel. All of those weird plasma waves and instabilities can transfer energy from one beam to another, or turn the plasma in the hohlraum into a mirror. By being careful about how energy is distributed across the 192 beams, NIF's scientists have managed to tame some of the laser-plasma interactions. Through experimentation, they discovered that reducing the density of the gas between the hohlraum and capsule a little, but not too much, allowed more of the laser beam energy to get into the gold. That meant more energy getting into the capsule, and more strongly driven implosions. Capsules are now regularly imploding at almost 400 km/s, more than thirty times Earth's escape velocity (even faster than NIF scientist and astronaut Dr. Jeff Wisoff traveled during space shuttle launches).

If NIF achieves ignition, it will be a sudden jump from brief spark to burning fusion plasma. As Mark Herrmann said, "In the simulations, it really is a cliff edge." While the gap from 3 percent to 100 percent may seem large, the improvements in energy yield that NIF has seen since 2011 have come in factors of between five or six. NIF's scientists aren't very many improvements away from net energy gain.

Omar Hurricane told me that plasma conditions were a much better measure than yield of energy alone of how close experiments are to ignition: "The metrics to focus on are actually the pressure, confinement time, or temperature, or you can also look at it as density times radius [of the hotspot within the fuel capsule]. The yield itself doesn't give a good impression of how close you are, as it's like approaching a cliff."

Different inertial fusion experiments are harder to compare with each other than magnetic fusion ones. The MagLIF experiment at Sandia National Laboratory has generated substantial numbers of neutrons using deuterium-deuterium reactions, but it isn't running with tritium.[16] Excitingly, First Light Fusion just started running with deuterium and tritium in 2020—but it hasn't yet published results on yield or the conditions it's reaching. All of which means NIF still tops the scoreboard for energy gain from inertial fusion.

While NIF has made great strides, to get that bit further requires more energy getting into the capsule's hotspot and kick-starting the fusion reactions. Each doubling of the energy dumped into the hotspot gives, roughly, ten times as much energy out. From the secret Halite-Centurion experiments, inertial star builders know that it is physically possible to get net energy gain if only they can build a laser big enough. Those experiments suggest that slamming a fusion fuel capsule with five to ten megajoules will result in net energy gain. But NIF is only fielding 1.8 megajoules right now.

So scientists at NIF are thinking about squeezing a few more joules out of their already phenomenally big laser. As Jeff Wisoff told me, it's expensive to add energy to the laser, so they have to try everything else first—but he said that a "modest upgrade" might be possible. Dr. Bruno Van Wonterghem, who spent years designing NIF and who took me on a tour of its steel arteries, told me about the improvements, reeling off what they'd done already, "but the ultimate is the bigger hammer"—that is, making the laser even bigger. Bruno smiled just a little when he told me that NIF had capacity for a 50 percent increase in laser energy. When I asked about the cost of a big laser upgrade, Livermore said that it would be a tiny fraction of the capital cost of NIF and Bruno gave the impression that it would be substantially less than their operating budget. If their sponsors at the Department of Energy go for it, he thinks they could do the work in a year or so. That might just be enough to get them to ignition.[17]

"When you look around the world," Jeff Wisoff tells me, "and say, 'Where is it credible we might get ignition in the next ten years?' I think NIF is the best shot." Unsurprisingly, NIF's director, Mark Herrmann, agrees, saying of ignition: "It needs a certain scale, energy, pressure; this is the only facility that has the prospect of doing that in the next decade."

The Race for Net Energy Gain Is Still Open

With machines like JET and NIF ever so close but each limited in some way, and ITER not even attempting net energy gain until 2035, there's a chance that another machine could swoop in and win the race.[18] Mark Herrmann told me that he spends a lot of time making sure NIF doesn't fall behind other countries' efforts. The international competition to achieve net energy gain looks set to become fiercer. Several other countries are planning, or have even built, laser fusion machines. They can learn from NIF—in particular, they know that because NIF has reached 3 percent, they probably don't need a laser that much bigger than 1.8 megajoules to hit 100 percent of net energy gain.

France's Laser MegaJoule was completed in 2014. I went to visit it in 2011, when it was under construction. As at NIF, a catacomb of thick, radiation-blocking walls had been constructed around the spherical target chamber. Because it hadn't yet been irradiated by the first experiments, I was able to poke my head into the shiny, echoey chamber and marvel at how it had been engineered to sub-millimeter precision and yet still was ready to withstand bombardment by millions upon millions of high-energy particles. Right now, Laser MegaJoule is fielding slightly less energy than NIF and it only works in the less-proven direct drive mode (which doesn't use a hohlraum), but it's one to watch.

Russian scientists revolutionized magnetic fusion when they revealed the tokamak. Now Russia has a plan to get out ahead on laser fusion. They're building a 192-beam laser facility, called UFL-2M. It will have 2.8 megajoules of energy, albeit in a less useful color than NIF's infrared beams. Russia also has a dark horse in the race: a fission-fusion hybrid reactor. Some star builders don't like fission-fusion hybrids because they lack many of the benefits of fusion alone. But some have passionately supported them as a more certain stepping stone that would bring net energy gain more easily within reach, and would allow for quicker development of fusion technologies.[19]

China is perhaps the most ambitious of all when it comes to fusion. "If you're prepared to do two or three different concepts and take a bit more risk, you can go faster," Ian Chapman told me. "That's what China is saying." If China follows through on its plans, it will be at the forefront of both tokamaks and laser fusion within the next decade, leaving Europe and the USA behind. Not only does it already have an existing 0.2-megajoule laser, Shenguang III, and EAST, a tokamak that reached temperatures of 100 million degrees Celsius (180 million degrees Fahrenheit) in 2018, it also has plans to build ignition scale devices that will rival or even surpass NIF and ITER respectively. "It's ambitious but not inconceivable," Ian said. "I think for China it's realistic because they're prepared to put the resources behind it."

It's an exciting period of nuclear innovation by governments that have deep pockets and rich nuclear expertise. For its part, Eurofusion, the club of European countries pursuing fusion, is also considering stellarators as a path to net energy gain—in particular, due to the astounding success of Wendelstein 7-X.[20]

And the private sector fusion firms claim that they're hard at the heels of the big laboratories, even if the data suggest they're behind right now. Most are promising to get to net energy gain (or, at least, net energy gain *conditions*) in the 2020s or early 2030s—well before ITER. General Fusion and Tokamak Energy see it happening by 2022, Lockheed Martin by the 2020s (revised from 2017), First Light Fusion by within the 2020s, and TAE, to stay on track, will need to have reached net energy gain *and* commercialized its technology by 2024. If we take these claims seriously, the star builders' race for net energy gain is close to finishing—but wide open.

Looking at the sweep of history, the progress made to date, and the promises being made from all quarters, it seems certain that net energy gain from fusion can and will be achieved. It could be this year. It could be next. Maybe it's thirty years away, like the joke about fusion says. For what it's worth, I think it will be sooner than that.

John Lawson showed us that net energy gain was theoretically possible. Extrapolation from magnetic star machines and interpolation from secret nuclear tests have shown us that it's experimentally possible. Just how fast net energy gain arrives will be determined by how badly nations and entrepreneurs want to see it happen.

But net energy gain *is* coming.

After Net Energy Gain

The latest star machines are tantalizingly close to demonstrating net energy gain from nuclear fusion reactions. So let's try a thought experiment: Imagine that net energy gain is achieved. Not just a small gain of one, but a big gain—thirty or more. A gain that is on the verge of what is needed to build a power plant. It could have happened with NIF, on one of the machines being built by a start-up, or maybe with ITER, when it's finished. For this thought experiment, it doesn't matter who. Now ask yourself a question: What *next*?

It's a question the star builders are increasingly asking themselves too: How will they take nuclear fusion from physics experiment to power plant?

As we sit in his office at Culham, Ian Chapman gives me a rundown of what needs to happen for tokamaks to successfully deliver energy to the grid. "There are five big challenges," he begins. The first is plasma that is ten times hotter than the center of the Sun: "JET is operating at 150 million degrees, so we know how to do that." This goal has been met, and any machine that gets net energy gain will have achieved it too. The temperatures used in inertial confinement fusion are typically lower than that, and NIF is just shy of what it needs for net energy gain. Ian lowers the thumb of his outstretched hand, four points to go.

"Second: How do we get the heat out of that big ball of gas with turbulence in the middle?" Once net energy gain is demonstrated, the energy produced must be extracted and used to turn water into steam. The steam will drive a turbine to produce electricity. Star builders have ideas about how to do this, but it's hard to make progress without a net-energy-gain reactor to try them on. For tokamaks, the first stage, getting heat energy out, is the hardest. "We sweep it down to a sacrificial surface at the bottom," Chapman continues. This is the divertor, a part of the tokamak designed to take a beating that's more intense than what the space shuttle receives during reentry. The bigger the tokamak, the more intense the heat. ITER is designed right at the physical limit of what solid materials in conventional tokamaks can endure. For inertial confinement fusion, the geometry of the reactor is much more simple and getting the heat out is too, so it's less of a concern. Ian lowers a finger, ticking exhaust heat off his list.

"The third challenge is neutrons: we [will] have the most intense source of neutrons on Earth." No one knows how surfaces will react to the number and energy of neutrons that a high-gain star machine will release—because building one is the only way to produce neutrons in sufficient numbers and with sufficient energy. Excessive neutron bombardment can make the hardiest materials crumble. Then there's the potential for excessive radioactivity if neutrons are absorbed by the wrong materials. Ian tells me that choosing the right materials for the reactor will make the difference between a plant that lasts four years and one that lasts forty. Due to the irradiation of the chamber by neutrons, repairs will be very hard to make, which could lead to reactors shutting down for long periods of time.

"The fourth challenge is breeding tritium." Tritium doesn't last long in nature because of its short half-life of twelve years. It has to be created from lithium. When lithium captures a neutron, it turns into tritium. Lithium is useful for many other items, like batteries, but Ian stresses that not much lithium is needed for tritium breeding because fusion has such a wildly high energy density. "For the rest of my lifetime," he says, meaning the energy he will use throughout his life, "I need a bathtub of water and the lithium you'd find in two laptop batteries." So the problem is not obtaining lithium, it's that no one's ever demonstrated breeding tritium from fusion neutrons and lithium on the scale needed—there's never been a reactor producing enough neutrons to try. After a few years of operation, the plan is that ITER will test out a lithium blanket around the reactor chamber, a first step in practical tritium breeding, but it won't be a production-grade system. Almost all digits down, Ian Chapman has just one finger in the air. We'll return to his fifth and final challenge for fusion later.[21]

Star builders today are thinking about how to conquer these challenges in a way that their predecessors just didn't need to. With net energy gain promised within fifteen years (not thirty!) by almost everyone, the need for solutions to the engineering challenges that will come with a fusion power plant is focusing minds.

Ian Chapman is taking the challenges very seriously indeed. Culham's current spherical tokamak, MAST Upgrade, is being used to explore the physics of the hot plasma exhaust that, in advanced tokamaks, will safely carry energy away from the hottest part of the plasma. They have a new Materials Research Facility tasked with looking at how materials will behave under the extreme conditions found in a nuclear fusion reactor, especially the constant bombardment by neutrons (there's a similar international facility funded by Japan and the EU). Also at Culham, a Fusion Technology Facility will try out different materials under all of the other stresses of fusion (electromagnetic, mechanical, and thermal) to see what works best. And there are tens of millions of pounds put aside for a Hydrogen-3 Advanced Technology Centre dedicated to the challenges of handling and breeding tritium. Similar facilities are being established by star builders across the world. Robotic arms are already being used for remote handling of irradiated materials on JET, helped by a new dedicated robotics facility that is trying to extend and improve the technology. Robots that can go in and fix problems with the reactor quickly can help prevent long periods of shutdown.[22]

It's one thing to overcome these challenges in isolation in specialist facilities, but it's quite another to achieve them together in a working, high-gain fusion reactor.

The magnetic fusion star builders are already planning to build another tokamak after ITER that will overcome the final barriers to commercialization and act as a demonstration power plant. This *will* deliver energy to the grid, and all that's necessary to achieve it: heat extraction (ultimately to drive a turbine), tritium breeding, robotic maintenance, and neutron-compatible materials. The power-to-the-grid reactor will be called DEMO, for demonstration power plant.

DEMO is more of an idea than a certainty, and the design is a work in progress. Star builders are waiting to see how ITER's first years of operation go before deciding what to keep and tweak in DEMO. What most designs suggest is that DEMO, if it gets the go-ahead, will generate five hundred megawatts, equivalent to a modestly sized nuclear fission power plant, and will have a gain of at least thirty times the energy put in.

DEMO faces many challenges. It probably needs to be slightly bigger than ITER; it needs to run with higher-density plasma; it needs to run for hours at a time, meaning that disruptions must be kept under control; and it needs to be completely self-sufficient in tritium.[23]

The laser star builders have thought about a demonstration power plant too. Inertial fusion faces more or less the same challenges as magnetic fusion, but one major difference is that inertial fusion reactors must rapidly repeat. Dr. Nick Hawker described this as First Light Fusion's biggest challenge—they'll need to have a shot as frequently as every five seconds.

NIF's laser currently manages four hundred shots per year. To be economical, laser fusion for energy will need to run with ten shots *per second*. The reason many repeats are required is that each capsule explosion on NIF doesn't release that much energy. For power generation, there needs to be a lot more energy coming out per second, and the energy needs to be released at a constant rate. Only by rapidly firing the laser at targets to produce mini-explosions, a bit like a gas engine in a car, can laser fusion devices hope to be commercially viable.[24]

With the current technology on NIF, this is impossible to do without breaking the laser. The optics need hours to cool down after each shot. Not to mention that the flashlamps take time to charge up and are horribly inefficient: of the 400 megajoules drawn from the grid, just half a percent (1.8 megajoules) currently make it into the laser beam.

Since NIF was designed several decades ago, laser technology has moved on, and the flashlamps that energize NIF's 192 laser beams have been surpassed by diode-energized lasers. The newer diode lasers are around twenty to forty times more efficient and produce less waste heat. But while they can fire an impressive ten times a second, they're a long way from being able to squeeze enough energy into each shot.[25]

Then there are the fiendishly complex targets that inertial fusion typically requires. For NIF, the cost of these needs to tumble from hundreds of thousands of dollars to just a couple of cents to make commercialization viable—remember that ten will be used up every second. When I raised with scientists at NIF how improbable that seems, they pointed out that modern manufacturing is effective at making precision-engineered products that cost cents—for example, bullets.

Livermore's plans for a power plant based on laser fusion found expression in a prototype design called LIFE (Laser Inertial Fusion Energy). The program to design LIFE was canceled when scientists couldn't get NIF to produce as much energy gain as they had predicted theoretically in the early 2010s. If NIF hits net energy gain, especially if it does so before any other star machine, then you can expect that the plans for an inertial fusion power plant will be quickly dusted off and scientists will focus on solving some of the fearsome challenges. But, for now at least, LIFE is dead.[26]

Notably, none of the challenges that face either the magnetic or laser approaches to fusion are true showstoppers: they're likely to be solvable with sufficient investment of resources and engineering skill. After all, until the current generation of star machines, no device had ever been the hottest place in the solar system or reached the density of the Sun's core.

The hard challenges of commercializing fusion and using it to save the planet don't end with engineering the reactor itself, however.

A major concern of star builders is that DEMO and LIFE, if they do happen, will come far too late for climate change. DEMO is scheduled for twenty years after ITER comes online—that is, in the 2050s—when many nations have said they'll already have achieved net zero carbon emissions. A LIFE-like reactor will probably only get into the planning stage after, and only if, NIF hits net energy gain. And neither LIFE nor DEMO will be *selling* electricity, even if they do supply it to the grid. They're demonstration power plants, not the first commercial fusion power stations.

The new fusion entrepreneurs are absolutely, painfully right that the pace of progress in fusion energy is going to be too slow to help save the planet unless star building gears up. Ian Chapman told me that fusion was better late than never. While that's true, if the star builders really want to ensure fusion energy is ready in time to be useful in preventing the worst effects of climate change, they need to step it up a notch.

Ian Chapman said that with the current funding and risk profile of the international collaborations on magnetic fusion energy, power wouldn't be getting to the grid until 2050. From that, he sees rapid deployment, "the same growth rate as fission had, about 30 percent year-on-year, and the same scale of plant and cost of capital." Even with this apparently rapid rate of deployment, starting from 2050 means that fusion would only power half the world's electricity needs by 2083. With few other options, policymakers may be forced to continue with a mix of energy production that includes fossil fuels, as well as nuclear fission and renewables.

The start-ups want to get fusion power onto the grid significantly more quickly. Tokamak Energy told me that star power must be deployed at "huge scales" *during* the 2030s. Nick Hawker has worked backward to get First Light Fusion's plan of action: "If we want to contribute to net zero by 2050 we need to be building plants, multiple, in the 2040s. And the first of a kind has to be built in the 2030s. Which means the physics problem has to be solved in the 2020s." If they're right, fusion plants will have to appear very quickly once net energy gain is achieved.

Can We Afford to Do Fusion?

The progress toward net energy gain, and more recently, the challenges that stand in the way of fusion power stations, all cost money. All of which means it's high time to find out what UK Atomic Energy Authority CEO Ian Chapman's fifth challenge is. "This one matters to me a lot," he says, more seriously. "If you do all of them, but the cost of electricity is chronic [prohibitive], then we have fusion, but it doesn't penetrate the market." All of the work, the battles against plasmas, the decades of innovation—they won't matter if the economics aren't right.

Money—to fund people, machines, or other resources—can speed up progress toward fusion. Conversely, lack of money can slow it down. When people ask how long it will be until fusion is achieved, they may as well be asking how much more money needs to be spent to get to a working reactor. The answer to how close fusion is as a power source depends on our collective appetite to see it happen.

Fusion is an expensive endeavor, partly because star machines seem to demand ever bigger scales—and bigger scales incur disproportionately bigger costs. ITER is going to be bigger than JET, DEMO is likely to be bigger yet. Initial estimates of the total cost of ITER were around €5 billion ($6.7 billion), but that had shot up to €20 billion (about $26.6 billion) by 2016. Even ITER insiders agree that scale is a problem. Sir Steve Cowley, the former CEO of the UK Atomic Energy Authority, has said, "I fully support ITER because we have to do a burning [self-sustaining fusion] plasma. But commercial reactors will need to be smaller and cheaper."[27]

NIF was supposed to cost $1.4 billion, but cost overruns took the total price tag to $4.1 billion.[28] Rough plans for a successor to NIF wouldn't need a bigger building, because of advances in laser technology, but would probably need to scale up the laser energy, and this adds to the cost. "That's why the NIF laser is not viable," First Light Fusion's Nick Hawker told me, when we were discussing commercialization. "The NIF laser was over $1000 per joule of energy, and our gain machine is going to be between four and five dollars per joule of energy. Even at that price, it's a problem."

Of course, first-of-a-kind reactors have research and development costs that don't apply to mature technologies. But star builders still think that capital costs will dominate commercial fusion reactors. "The actual fuel costs will be small," Professor Sibylle

Günter told me. She listed the two biggest costs as the initial investment for the device and buildings, and the maintenance of the plant. This could be a problem, as finding funding for capital intensive power plants, with payoffs over decades, isn't easy. The UK government, for example, has struggled to find private financing for a mammoth three-thousand-megawatt fission plant at Hinkley Point in Somerset.

Here's where the *new* star builders have a killer argument: if their machines work, they'll be smaller and less capital intensive. "A small, modular device means more can be built off-site in a factory, and that's a big factor to driving cost down," Tokamak Energy's CEO, Jonathan Carling, told me. Tokamak Energy has published academic research suggesting that the need for scale in tokamaks may be a false supposition. They hope that their spherical tokamaks can sidestep the need for ever bigger machines. The huge price drops that can be achieved with small modular devices is a well-known phenomenon—Moore's Law, for microchips, is one famous example. Solar cells are another: they became more than three thousand times cheaper between 1956 and 2019. First Light Fusion is keeping costs down by relying on off-the-shelf technologies whenever possible.[29]

However, it's a big "if." What if we have to rely on government attempts to do fusion, which so far have utilized huge machines? There's no doubt that the price tags for JET ($2.6 billion), NIF ($4.1 billion), and ITER ($22 billion) are high, but maybe that's not intolerable. Those sums aren't so different from the cost of the Large Hadron Collider at CERN ($5.3 billion), the Square Kilometre Array ($1 billion), and the most expensive big science project ever, the International Space Station ($120 billion). The next generation of particle collider will probably cost $25 billion. Even with their high price tags, these are all extremely worthwhile programs that will enhance our species' existence. But while all will deliver new scientific knowledge, star builders would say that only the fusion machines are capable of delivering a new energy source. And a new energy source is desperately needed.[30]

When the costs of existing experimental reactors are spread over the world's population, they're minor. The going rate of around $10 billion is about the same as a US aircraft carrier. The first commercial reactor would be in the same ballpark, with subsequent reactors much cheaper. Compared to the USA's total spend on energy of more than $1,000 billion each year, the cost of fusion reactors is small.[31] It's positively tiny compared to the cost of America's war in Iraq, which is in the multiple thousands of billions.

But these are somewhat silly comparisons. It's more sensible to ask whether funding for fusion is enough for it to happen on a reasonable timescale. It hasn't been, at least historically—and we've known that for a long time.

In the 1970s, the US Energy Research and Development Administration put together estimates of how quickly fusion might be achieved under different levels of annual funding. They estimated that the minimum level of funding that would see commercialization of fusion ever happen was $2 billion per year. From 1976 to 2012, the average US level of funding for fusion energy sciences was roughly $0.6 billion per year for magnetic and laser fusion combined. That's well under the minimum that it was thought might be required to ever reach a prototype fusion reactor, whether via tokamaks or giant lasers.[32]

Funding of around $0.6 billion per annum (which has also been the allocation in more recent years) may sound like a lot, but in 2018 it was only 0.1 percent of the US's research and development budget and less than what big oil and gas firms spend individually per year on research and development.[33]

Worldwide, by far the largest fusion expenditure is on ITER, which has totaled around $1.7 billion per year over its long gestation period. Again, that sounds like a lot, but it's only around a tenth of 1 percent of the global spend on research and development.

Moreover, it's good to remember that 9 million people a year are dying from air pollution and 150,000 or so are already dying from climate change–related phenomena such as drought, floods, and fires. Fusion comes with risks because—despite what some star builders say—it's impossible to know for certain whether it will ever be commercially viable. But the potential upsides of eliminating those unnecessary deaths seem worth it.

For context, consider the fact that the world spends roughly $20 billion per year on research into cancer, which is thought to be responsible for 10 million deaths a year.[34] Cancer research today can probably translate into lives saved or improved much more quickly than fusion research (because the latter will only be helpful if it results in technology that is commercially viable), so it's certainly not a perfect comparison. What the numbers do highlight, though, is that fusion is currently less well funded than you might expect given its potential.

People who complain that fusion is taking a long time should be aware that money is a large part of the reason. Investment is key to speeding up or slowing down progress. And we've known since the 1970s that investment levels in fusion are too low to make rapid progress. It's not that net energy gain isn't *possible*—it's just that we haven't wanted it badly enough. If we really want scientists to save the planet by building a star, someone's going to have to pay for it.

The funding situation is changing. "Five years ago no one in government spoke about fusion," Ian Chapman told me. "Now, when I go into the Treasury, people want to have discussions about it." As he also noted, China has a much bigger appetite for risk and funding when it comes to fusion, and their plans suggest they may set the future pace of global research. As for the money from private investors that is now pouring into fusion, some of it is being directed toward projects with very long odds, but as star builder David Kingham put it, private sector fusion grows the pie. To make fusion deliver in time, it's *all* going to be needed.

At the risk of getting ahead of ourselves, it's interesting to speculate on how expensive fusion energy will be for consumers if and when it's commercialized. To spotlight again Ian Chapman's fifth challenge, an energy source that is perfect in every other way but is one hundred times more expensive than anything else isn't going to achieve much in the way of market penetration. To compare the costs of different types of power, the LCOE (levelized cost of electricity) is often used. It's calculated by comparing the costs of a power plant with the entire amount of energy that the plant will produce over its lifetime. It's usually measured in dollars per amount of energy produced. For instance, based on a number of countries, the International Energy Agency estimates that a conventional coal power plant has an estimated LCOE of $24 per gigajoule (a gigajoule is a thousand megajoules). It's a useful way to think about how expensive energy sources are.

It's hard to know how expensive fusion energy will be because there are so many unanswered questions. Will fusion plants have different regulatory requirements from fission plants? How big will a plant be? How long will a reactor chamber last? How much downtime will there be? Any estimate of the cost of fusion has to make heroic assumptions to get to a number. A recent US Department of Energy report on the likely costs of nuclear technologies refrained from calculating an LCOE precisely because of this uncertainty— and even the LCOE values for established technologies are significantly different depending on the assumptions used. That's not to say people haven't tried, but any estimates should be taken with a grain of salt.

One of the best guides to the likely cost of fusion power is the cost of fission power. Fission power plants are a good comparison because they're also a nuclear technology and would probably need a similar scale of facility—though the regulations for fusion are likely to be significantly less stringent than they are for fission. The cost of fission power is an upper bound for fusion given that the overall costs of fission include dealing with long-lived radioactive waste and risks like meltdown. What is similar between fission and fusion is that much of the expense is likely to be in the form of capital costs covering construction, commissioning, and financing (the fuel only makes up a small part of the LCOE for nuclear technologies). The International Energy Agency put the price of advanced nuclear fission reactors at $19 per gigajoule, cheaper than conventional coal.

As part of the designs for the prototype fusion power plants, DEMO for magnetic fusion and LIFE for inertial fusion, cost-of-electricity estimates were made. For a successor to DEMO, cost is reckoned at $21 to $45 per gigajoule. LIFE came in a bit lower, on the high end at $21 to $30 per gigajoule. Nick Hawker has published detailed estimates of cost ranges for his more modular machine that suggest $30 per gigajoule is likely but that as low as $7 per gigajoule could be possible.[35] Twenty to thirty dollars per gigajoule isn't too far away from the estimates for fission, so seems plausible. But, honestly, no one really knows. What we can say is that it's unlikely that fusion will have an LCOE that is dramatically less than the LCOE of nuclear fission, *unless* star builders find a way to substantially shrink the size of the plant required, via new technologies such as smaller lasers or better superconducting magnets.[36]

What about other power sources?* The cheapest fossil fuel is the combined gas cycle turbine at $18 per gigajoule, but that cost

---

*The quoted figures are medians for a range of countries as estimated by the International Energy Agency with a discount rate of 7 percent. The order of prices given is similar to the order of prices for electricity given by the US Energy Information Administration.

doesn't take into account the negative externalities of air pollution and carbon dioxide. The other fossil fuels are relatively expensive. The really interesting comparisons are with renewables; concentrating solar thermal (using mirrors to heat a liquid) is $34 per gigajoule, industrial-scale solar photovoltaic (panels) is $16, offshore wind is $24, and onshore wind is $14. Solar cells and onshore wind are already the most competitive forms of power generation, and fusion is unlikely to beat them on price.[37]

Fusion isn't magic and it certainly won't be free or too cheap to meter.[38] It's like Nick Hawker said: the problem with power generation isn't cost, because wind and solar are already the cheapest power sources, and are likely to remain so. If we take the cost estimates seriously, then the price of fusion energy will be competitive, perhaps one of the lowest, but it's unlikely to be *the* cheapest form of power.

What fusion can buy the world is carbon-free energy on the scale that we need, and at the rate of deployment that we need, for the period of time that we need, in order to save the planet.

CAN WE AFFORD NOT TO DO FUSION?

> "Thermonuclear energy will be ready when it becomes necessary for humanity."
>
> —Lev Artsimovich, head of the Soviet Fusion Program, 1951–1973[1]

Life on Earth is robust but fragile. It's robust because life has been going on for at least 3 billion years. It's fragile because most species that have ever existed have gone extinct. When I think about the planet's most successful species, I don't think about humans. I think about crocodiles and coelacanths.

Crocodiles have racked up 85 million years as a species. But even hardy crocodiles are novices compared to another species that has persisted for *400 million* years. The coelacanth is a lobe-finned fish species that was thought to have died out 66 million years ago. Apparently, no one told the coelacanths this because, in the 1930s, one was caught by a fishing boat. Far from going extinct, they've survived four of Earth's five mass extinctions over their incredible run (you can see the 66-million-year-old fossil of a coelacanth next to a twentieth-century specimen preserved in formaldehyde in London's Natural History Museum). In our current form as *Homo sapiens*, we've only been around for a few hundred thousand years.

We need to start playing the long game when it comes to our own survival. We need to be more coelacanth.[2]

Mass extinctions and global catastrophes do happen. All too recently, we've seen how a pandemic can leave death and economic ruin in its wake, and how unpreparedness can amplify its worst effects. But the Earth as a whole isn't safe from other terrors. Three of the biggest existential threats we face are asteroid or comet impact; massive volcanic eruptions, called super-eruptions; and—unique to us—runaway human-caused climate change.

An asteroid about eight miles wide is probably responsible for wiping out the dinosaurs and 75 percent of all other species at the time. The dinosaurs reigned for more than 100 million years, so they weren't too shoddy in the survival stakes either. In 1908, we got a tiny, frightening taste of this kind of event when an asteroid that was at most a few hundred meters wide disintegrated above Siberia and destroyed more than 2,000 square kilometers (approximately 772 square miles) of forest. The chances of an asteroid or comet big enough to kill most of the world's population smashing into Earth are around one in ten thousand per century. Super-eruptions are more likely and would affect at least half the planet and probably all of it, raining down ash over entire continents.[3]

The aftermath of these extinction-scale events would be enormous climate change. Dust and earth from asteroid impacts or volcanic eruptions could block out part of the Sun's light, producing a cooling effect and making it harder to grow food or use solar energy.

What can we do to prepare for these rare but world-changing disasters? Having a source of energy that can keep going despite sweeping and adverse changes in climate seems like a good precaution. As we know, fossil fuels will run out before too long. And renewables that rely on large areas of land are susceptible to environmental changes. Fission could be one solution. Star power is another: the fuels are (relatively) common—deuterium is found in all the world's oceans while lithium is found on all the world's inhabited continents—and in any case not that much of either is needed.

This may all sound scary. It is. But it's also prudent to think about in the long run. We *know* that these events can happen. On long enough timescales, they're almost certain to. Personally, I'd like humanity to thrive far into the future. The choices we make today have enormous consequences for future generations. Star builders say that using a fraction of world research budgets to perfect fusion energy is a small price to pay for a disaster-resilient power source.

Of course, there are reasons beyond just saving our skin to want to see fusion achieved.

The pursuit of fusion has led to scientific discoveries that are among the most extreme and surprising of any field. Think for a second about plasmas. Understanding them is key for fusion, yes, but every plasma discovery also gives us a better understanding of 99 percent of the visible universe. Plasmas are one of the most dramatic examples of how "more is different": while we might understand the behaviors of the individual components—nuclei and electrons—something changes when they're combined in large numbers and complexity emerges out of simplicity.[4] Understanding these rich phenomena can be its own reward, as with so many other topics in science. Even if the study of plasmas wasn't worth it for the joy of discovery alone, their emergent complexity holds practical lessons for other subjects, like economics, where people's interactions are also different from the sum of their parts.[*5]

A growing understanding of plasmas has led to more practical applications too, like cleaning surgical equipment or, quite literally, growing diamonds.[6] Also, using lasers and plasmas together

*One of the equations at the heart of plasma physics, the Fokker-Planck equation, has recently made its way into macroeconomic models.

has resulted in new and better ways to fight cancer: lasers can be used to accelerate protons in a plasma to very high energies, and those protons can more accurately target cancerous cells than, for example, X-rays.[7]

Machines like NIF are doing incredible science in addition to the experiments on inertial fusion energy and stockpile stewardship. NIF has been used to re-create the conditions in the core of stars that have ten times the mass of the Sun, leading to better estimates of their rate of fusion reactions.[8] NIF experiments have taken us elsewhere in space too. Shortly before his death at the age of ninety-five, Livermore founder Edward Teller told scientists there that what he wanted for his one hundredth birthday was to get "excellent predictions—calculations and experiments—about the interiors of the planets."[9] The strides forward came too late for Teller, who passed away in 2003, but since NIF opened, Livermore's scientists have managed to re-create the huge pressures of the interiors of gas giants like Jupiter and Saturn, albeit on a tiny scale. In the experiments, NIF's laser beams were used to compress liquid deuterium to 6 million times Earth pressure and to a temperature of a few thousand degrees. As the pressure increased, usually transparent deuterium liquid first became opaque and then, most remarkably and bizarrely of all, turned into a shiny metal. It is like squeezing your coffee cup and finding that it has turned into a plate.[10]

The engineering challenges that star builders like Professor Ian Chapman are solving en route to commercializing fusion have industrial spin-offs. Culham's remote handling robots can be used in many situations that require dexterity but aren't safe for humans to enter. Also, developing resilient fusion reactors is pushing engineers to create new materials suitable for extremes.[11] Lawrence Livermore has filed a host of patents based on what they've had to invent to make NIF work. As happened with research into crewed space flight, fusion research is driving innovation beyond its own needs.

The scientific and industrial reasons to pursue fusion, good as they are, aren't the most bold or ambitious arguments to perfect the power source of stars though. There's a reason to achieve fusion that speaks even more loudly to our existence as a species—which is that we could spread our wings and explore the universe.

Venturing farther into space sounds like a wild dream, and it is, for now. But we've been to the Moon. We've landed an un-crewed spacecraft on an asteroid. We've sent probes outside of the solar system. Before too long, we may send a crewed mission to Mars. What person doesn't want to open the next door and see what's waiting for us in the rest of the universe?

The only way we'll travel to the universe beyond our celestial backyard is with plasma physics and nuclear fusion. Fusion rockets are humanity's best hope for traveling across the vast distances of space.

Rocket science has a reputation for being complicated, but it all boils down to two simple ideas. The first is this: if you expel stuff in one direction, you'll travel in the other direction. The second relates to how much force can be created. That's determined by how much mass is being expelled, and how quickly it's being expelled. Rockets expel a lot of mass quickly to get into orbit. But there's a problem with expelling lots of mass; you have to carry the mass with you until it's expelled. The more force you need, the more mass you have to carry, which increases the force you need, and so on. To avoid this problem, propulsion methods that work in space won't be able to rely on expelling lots of mass; instead they'll need to maximize the speed of the mass being expelled (and to expel only a little mass).

You can probably guess why fusion is a good choice for the hasty space traveler. Fusion can create enormous exhaust speeds with only small amounts of mass because of its high energy density. While the best chemical rockets can achieve exhaust speeds of 4.5 kilometers (about 2.8 miles) per second, and nuclear fission might reach 8.5 kilometers per second, a working nuclear fusion reactor could potentially produce exhaust speeds of hundreds to thousands of kilometers per second.[12]

Sticking a fusion reactor on a spacecraft is, surprisingly, not the only fusion-spacecraft option out there. Project Orion was part of Edward Teller's "Plowshare" program to turn nuclear weapons to peaceful purposes* and was co-led by physicist Freeman Dyson.[13] It looked at chucking exploding hydrogen bombs out of the back end of a spacecraft to cause it to accelerate in the other direction. The scheme isn't quite as insane as it may seem, and Dyson himself estimated that it could produce exhaust speeds of one thousand to ten thousand kilometers (approximately six hundred to six thousand miles) per second. Apart from this approach to fusion-powered space travel posing significant proliferation and safety risks, tests of pulsed nuclear explosion rockets are effectively banned by international treaties—so it seems much more sensible to use controlled fusion reactors to achieve similar ends. However fusion-powered rockets are ultimately achieved, research into fusion for energy will aid their development.[14]

Anyway, developing fusion propulsion isn't just about exploring the universe; fusion rockets could also help us prevent planetary-scale extinction of life from happening in the first place. The major challenge in preventing a humanity-killing asteroid or comet is detecting and reaching it early enough so that mitigating action—such as steering it out of the way—can be taken. Giving an asteroid a small push early on is as effective as a big push later on. Fusion-powered rockets could travel through space faster than conventional rockets, buying us more time to take action. And in

---

*Named after the passage in the Bible: "and they shall beat their swords into plowshares." Isaiah 2:3–4.

the catastrophic event that we couldn't save the Earth, the ability to travel to a new home would be the ultimate insurance policy for humanity.

Nuclear fusion reactor–powered spacecraft bring space travel within reach. They'd cut down the time it takes to get to Mars significantly, making it possible to do round-trips within a year. They could even allow us to travel outside of the solar system. The closest star system outside of our own is centered on Proxima Centauri, a red dwarf star four light-years away—that is, it takes light four years to make the journey from Proxima Centauri to Earth. Proxima Centauri has a habitable zone, a region where—in principle—life could exist. Within that habitable zone sits Proxima Centauri B, the closest exo-planet to Earth. While it's highly unlikely that Proxima Centauri B is habitable, we don't yet know. With a fusion rocket, a trip to Proxima Centauri B would be possible in under forty years, a remarkably short time.[15]

I began this book with a crazy idea—to create a slice of star matter and do nuclear fusion reactions in it to produce energy. The scientists, entrepreneurs, and governments that have pursued this goal aren't so crazy though. The scientists are some of the best there are. Some whom we've met are trusted with stewardship of the United States' nuclear arsenal; they really know what they're doing when it comes to nuclear physics. Others, like Professor Ian Chapman at Culham, have won multiple awards for the quality of their research. The entrepreneurs in the race to build a star are daring and ambitious, raising amounts that most start-ups can only dream of and making decades' worth of progress in fusion in just a few years. The governments funding fusion are leading the richest countries, and represent a majority of the world's population.

Their motivations don't seem crazy either. We're causing an unprecedented change in our environment, and it's mostly driven by our use of energy. But our use of energy has improved life in ways that would have seemed impossible to our forebears. Reducing energy demand enough to stop climate change seems impractical; if anything, we're likely to need more energy, not less. We have technologies to get us partway there—most notably, solar and wind power, and fission too, where it's accepted. But they won't get us the whole way. Those pursuing this apparently crazy idea say that we *can* have it both ways: we can improve quality of life for many more people *and* protect the environment at the same time. Fusion could deliver CO_2-free energy at scale and is likely to be one of the safest power sources, if not *the* safest, ever devised. Climate change aside, we need new sources of energy because our primary sources of energy, fossil fuels, are dwindling. The ingredients of even the most basic form of fusion, with deuterium and tritium, could last us around 33 million years. It's not as long as the coelacanths have been swimming around, but it's a damn good start. And those 33 million years would surely buy us enough time to figure out how to do fusion reactions that use even longer-lasting fuel.

Even if you believe that fusion could save the planet, you might not be convinced that it's possible to make it work. And yet nature tells us not only that fusion can happen, but that it's by far the universe's most ubiquitous power source—the one that lights each and every day on Earth, and maps the heavens at night. The universe's visible matter began with nuclear fusion, and stars end with it when they go supernova. We couldn't exist without the atoms in our bodies being forged by it. Fusion is everywhere, so, star builders say, why not on Earth too? As a species, we've already harnessed controlled fission (in today's nuclear plants), and both uncontrolled fission and fusion in nuclear weapons. Is it really so crazy to think that we might be able to harness controlled fusion too?

Star builders say "no!" and the machines that they've built have come very close to demonstrating that scientifically with net energy gain. Magnetic confinement fusion has reached 67 percent of gain in fusion power, inertial confinement fusion 3 percent in energy. More important, we know from pen-and-paper physics that net energy gain from fusion is possible. Experimental evidence strongly suggests that inertial confinement fusion *will* produce net energy gain with a big enough laser. The conditions for self-sustaining nuclear fusion, or ignition, are close, and there have been millionfold improvements toward them since the first fusion machines were built. That progress has taken a long time, and now entrepreneurs are challenging the slow-moving government laboratories, pushing progress to become faster, cheaper, and more commercially viable. While there is not complete agreement on *how*, and the *when* will depend on money and luck, star builders all say that net energy gain *is* coming.

Star builders are now looking *beyond* net energy gain, and setting their sights on putting fusion energy on the grid. Huge engineering and commercial challenges remain. Entire plants must first produce more energy than they use (every day, and not just in individual experiments). Fusion energy must be extracted and turned into electricity in a safe and sustainable way. And fusion energy must be both widely available and affordable. To go from no fusion power to significant fusion power is a change on a scale that is difficult to appreciate, requiring thousands of plants to be constructed all over the world. But many star builders won't think they've really succeeded unless fusion is delivered quickly enough, and on big enough scales, to help the planet avert a climate catastrophe that—right now—is coming toward us like a juggernaut.

It's been a long wait to get even this far in the race to power the planet, but it's worth asking two questions: Is progress being made? And is the destination worth it? My star-building adventure has convinced me that the answer to both of those questions is absolutely yes. Given the potential benefits, it seems clear that we can't afford *not* to do fusion.

And if we do forge ahead, it seems highly likely that, many centuries from now, if we're still here at all, we'll be enjoying clean energy—indirectly, from solar power, and directly, from star machines.

Just how close that fusion-powered future is will depend on how much we want—no, *need*—it to work. And the people I've met on this adventure say that, ultimately, we *will* need it: because fusion is the only power source that can take us to the stars.

ACKNOWLEDGMENTS

So many people were generous with their time and encouragement while I was writing this book, and I couldn't have done it without them.

I want to begin by saying a special thank you to Melanie Windridge, who has helped at every stage. If Melanie hadn't invited me to her own book launch for *Aurora*, I'm not sure I would be writing this at all. And it was Diane Banks of Northbank Talent Management who came up to me at that launch and gave me the confidence to think people might be interested in the story I had to tell—thanks for believing in me from the first moment, Diane.

My agent at Northbank, Martin Redfern, has been an indispensable source of information on the weird workings of the publishing world and fought hard to land this book with excellent editors (which he did). I couldn't have made the book without him, and I'd like to particularly thank him for his endless patience with my many questions. Thanks also to my former agent, Robyn Drury, (now a commissioning editor at Penguin) who somehow saw the potential of this work even in that very first draft. She also provided sage advice on how to make a book about nuclear and plasma physics into something people might actually want to read!

I want to give a special thanks to those who provided feedback on individual chapters: Brian Appelbe, Stephen Pugh, Paul Robinson, Steve Rose, and Mark Sherlock. Of course, it's because of Mark and Steve, my PhD advisors, that I got into plasma physics and fusion in the first place. Thanks also to those I discussed bits of the book with: Charlotte Palmer, Jerry Chittenden, Ed Hill, Andrew Holland, Matthew Lilley, Stuart Mangles, and, of course, Oli Pike, who first encouraged me to write a book on fusion.

My colleagues at the Bank of England were enormously supportive while I was writing this book—I'd like to thank David Bholat in particular for his encouragement and enthusiasm.

Thank you to everyone in academia, industry, and government who generously gave their time to help with the book in one way or another: David Kingham, Steve McNamara, Jonathan Carling, Lorne Horton, Fernanda Rimini, the stars of the *A Glass of Seawater* podcast and other CCFE PhD students (sorry that I didn't get everyone's names!), Ian Chapman, Howard Wilson, Dave Stephens, Chris Warrick, Karl Tischler, James Pecover, Guy Burdiak, Nick Hawker, Gianluca Pisanello, Nathan Joiner, Hugo Doyle, Isabella Milch, Sibylle Günter, Emma Chapman, Jeff Wisoff, Michael Stadermann, Becky Butlin, Mark Herrmann, Tayyab Suratawala, Brian Welday, Bruno Van Wonterghem, Louisa Pickworth, George Swadling, Omar Hurricane, Steve Cowley, and Jason Parisi. A special thank you to Breanna Bishop at Livermore and Nick Holloway at Culham for scheduling my visits.

Thank you to Laurie Winkless for the advice on (science) writing.

A big thank you to the teams at Weidenfeld & Nicolson and Simon & Schuster. Clarissa Sutherland and Beckett Rueda provided help that seemed well beyond the call of duty. I was extremely lucky to have such eagle-eyed copyeditors as Jo Gledhill and Rick Willett. And I was bowled over by the cover designs by Jason Anscomb, for the UK and Commonwealth, and Jonathan Bush, for the US. I want to thank Paul Murphy, my first editor at Weidenfeld & Nicolson, who immediately shared my vision for this book; I hope you're proud of what we've done with it, Paul. Most of the book has come together under the remarkable aegis of my two editors, Maddy Price at Weidenfeld & Nicolson and Rick Horgan at Simon

& Schuster. They are a dream team. Maddy has been phenomenally generous with advice and I'm not sure there's a single comment of hers I didn't ultimately put through because each and every one made the book better. Rick is quite simply a force of nature and, at every turn, has pushed me to do better for my readers and for myself as an author. I'm so grateful for the opportunity I've had to work with, and learn from, both of them.

Finally, my biggest thanks go to my wife, Alice Turrell, who read almost *every* chapter of *every single draft*, and was indefatigable in providing critique and encouragement; both have been invaluable.

NOTES

Prologue—A Crazy Idea

1. Bill Gates, "Two superpowers we wish we had—Annual Letter 2016" (2016), https://www.gatesnotes.com/2016-Annual-Letter.
2. Lawrence Livermore National Laboratory, National Ignition Facility and Photon Science (2019), https://lasers.llnl.gov.
3. S. Atezni and J. Meyer-ter-Vehn, *The Physics of Inertial Fusion* (Oxford Science Publications, 2004).
4. "The Boy Who Played with Fusion," *Popular Science* (2012), http://www.popsci.com/science/article/2012-02/boy-who-played-fusion; "Young Scientist Jamie Edwards in Atomic Fusion Record," BBC News (2014), https://www.bbc.co.uk/news/av/science-environment-26450494/young-scientist-jamie-edwards-in-atomic-fusion-record.
5. "A Tool for Tracking Millions of Parts," *Iter Newsline* (2014), https://www.iter.org/newsline/-/1887; *Space Shuttle Era Facts*, NASA (2011), https://www.nasa.gov/pdf/566250main_2011.07.05%20SHUTTLE%20ERA%20FACTS.pdf.
6. "Stephen Hawking: Why We Should Embrace Fusion Power," BBC News (2016), https://www.bbc.com/future/article/20161117-stephen-hawking-why-we-should-embrace-fusion-power.
7. "Boris Johnson Jokes About UK Being on the Verge of Nuclear Fusion," *New Scientist* (2019), https://www.newscientist.com/article/2218570-boris-johnson-jokes-about-uk-being-on-the-verge-of-nuclear-fusion/#ixzz66tYUwh6k.
8. R. F. Post, "Controlled Fusion Research—An Application of the Physics of High Temperature Plasmas," *Reviews of Modern Physics* 28 (1956): 338.
9. R. Herman, *Fusion: The Search for Endless Energy* (Cambridge University Press, 1990).
10. S. Cowley, "Fusion Is Energy's Future," TED Talk (2009).
11. "FOCUS FUSION: emPOWERtheWORLD," IndieGoGo (2014), https://www.indiegogo.com/projects/focus-fusion-empowertheworld--3\#.
12. J. Tirone, "Nuclear Fusion," *Bloomberg* (2019), https://www.washingtonpost.com/business/energy/nuclear-fusion/2019/06/20/c6bd5682-938d-11e9-956a-88c291ab5c38_story.html.
13. "PayPal Billionaire Peter Thiel 'Becoming Key Donald Trump Adviser,'" *Independent* (2017), http://www.independent.co.uk/news/world/americas/us-politics/peter-thiel-donald-trump-key-adviser-technology-science-paypal-david-gelertner-steve-bannon-a7600471.html; "Peter Thiel's Other Hobby Is

Nuclear Fusion," *Bloomberg* (2016), https://www.bloomberg.com/news/articles/2016-11-22/peter-thiel-s-other-hobby-is-nuclear-fusion.

14. "The Secretive, Billionaire-Backed Plans to Harness Fusion," BBC News (2016), https://www.bbc.com/future/article/20160428-the-secretive-billionaire-backed-plans-to-harness-fusion.

15. "Oil Major Chevron Invests in Nuclear Fusion Startup Zap Energy," Reuters (2020), https://www.reuters.com/article/us-chevron-investment-nuclear-idUS KCN25831E; "The Secret U.S.–Russian Nuclear Fusion Project," ZDNet (2013), http://www.zdnet.com/article/the-secret-us-russian-nuclear-fusion-project/; R. Martin, "Go Inside TriAlpha, a Startup Pursuing the Ideal Power Source," *MIT Technology Review* (2016), https://www.technologyreview.com/s/601 482/go-inside-trialpha-a-startup-pursuing-the-ideal-power-source/; "Lockheed Portable Fusion Project Still Making Progress," *Next Big Future* (2016), http://www.nextbigfuture.com/2016/05/lockheed-portable-fusion-proejct-still.html; "The British Reality TV Star Building a Fusion Reactor," BBC News (2017), https://www.bbc.com/future/article/20170418-the-made-in-chelsea-star-building-a-fusion-reactor.

16. T. Peckinpaugh, M. O'Neill, and A. Johns, "U.S. House of Representatives Demonstrates Support for Fusion Energy," https://www.globalpowerlawand policy.com/2020/09/u-s-house-of-representatives-demonstrates-support-for-fusion-energy/ (2020).

17. Comments given at Emergence of the Fusion Industry Breakfast organized by the Fusion Industry Association, Wednesday, March 4, 2020, in London.

18. Max-Planck-Gesellschaft, "Angela Merkel Switches on Wendelstein 7-X Fusion Device," https://www.mpg.de/9926419/wendelstein7x-start (2016).

19. M. Herrmann, "The Future of U.S. Fusion Energy Research—Hearing: Delivered to the Committee on Science, Space, and Technology Subcommittee on Energy" (2018).

20. "The Secret U.S.–Russian Nuclear Fusion Project, ZDNet (2013), http://www.zdnet.com/article/the-secret-us-russian-nuclear-fusion-project/; International Atomic Energy Agency, Fusion Device Information System, https://nucleus.iaea.org/sites/fusionportal/Pages/FusDIS.aspx (2020); "China Plans Fusion Power Research," *World Nuclear News* (2019), http://world-nuclear-news.org/Articles/China-plans-fusion-power-research; Pravda. ru, "Russia Prepares to Test Laser More Powerful Than USA's National Ignition Facility," YouTube, https://www.youtube.com/watch?v=uTqg-tmD dEg; "Will China Beat the World to Nuclear Fusion and Clean Energy?," BBC News (2018), https://www.bbc.co.uk/news/blogs-china-blog-43792655; "China Targets Nuclear Fusion Power Generation by 2040," *Euronews* (2019), https://www.scmp.com/news/china/science/article/2177652/operation-z-machine-chinas-next-big-weapon-nuclear-arms-race.

21. "Billionaires Back Fusion Energy Projects in Pursuit of a SpaceX Moment," *Seattle Times* (2018), https://www.seattletimes.com/business/billionaires-back-fusion-energy-projects-in-pursuit-of-a-spacex-moment/.

Chapter 1: The Star Builders

1. A. S. Eddington, "The Internal Constitution of the Stars," *Observatory* 43 (1920): 341–58.
2. I. T. Chapman, "Modelling the Stability of the N=1 Internal Kink Mode in Tokamak Plasmas" (Imperial College London, 2008).
3. "The Joint European Torus Is Going Out with a Bang," *Science Business* (2019), https://sciencebusiness.net/news/joint-european-torus-going-out-bang.
4. J. Bairstow, "Tokamak Energy Wins $580k from US Government to Tackle Fusion Challenges," *Energy Live News* (2020), https://www.energylivenews.com/2020/09/07/tokamak-energy-wins-580k-from-us-government-to-tackle-fusion-challenges/.
5. "Oxford Startup Promises Fusion Gain by 2024," *EENews Europe* (2019), https://www.eenewseurope.com/news/oxford-startup-promises-fusion-gain-2024#.

Chapter 2: Build a Star, Save the Planet

1. "Gods of Science: Stephen Hawking and Brian Cox Discuss Mind Over Matter," *Guardian* (2010), https://www.theguardian.com/science/2010/sep/11/science-stephen-hawking-brian-cox.
2. Energy and Climate Intelligence Unit, "One-Sixth of Global Economy Under Net Zero Targets" (2019), https://eciu.net/news-and-events/press-releases/2019/one-sixth-of-global-economy-under-net-zero-targets.
3. R. N. Carmody and R. W. Wrangham, "The Energetic Significance of Cooking," *Journal of Human Evolution* 57 (2009): 379–91; R. Wrangham and N. Conklin-Brittain, "Cooking as a Biological Trait," *Comparative Biochemistry and Physiology Part A: Molecular & Integrative Physiology* 136 (2003): 35–46.
4. R. J. Gordon, *The Rise and Fall of American Growth: The US Standard of Living Since the Civil War* (Princeton University Press, 2016).
5. BP, *Statistical Review of World Energy 2020* (British Petroleum, 2020); UK Government, *Digest of United Kingdom Energy Statistics 2020* (UK Department of Business, Energy and Industrial Strategy, 2020); R. Fouquet, *Heat, Power and Light: Revolutions in Energy Services* (Cheltenham, UK: Edward Elgar Publishing Limited, 2008); R. Fouquet, "Consumer Surplus from Energy Transitions, *Energy Journal* 39 (2018); V. Smil, *Energy Transitions: History, Requirements, Prospects* (Westport, CT: Praeger, 2010); H. Ritchie, *Energy. Our World in Data* (2014), https://ourworldindata.org/energy.
6. T. Cowan, "Want to Help Fight Climate Change? Have More Children," *Bloomberg* (2019), https://www.bloomberg.com/opinion/articles/2019-03-14/want-to-help-fight-climate-change-have-more-children; M. Kremer, "Population Growth and Technological Change: One Million BC to 1990," *Quarterly Journal of Economics* 108 (1993): 681–716.
7. World Bank, *Total Population. World Bank Indicators* (2019), https://data.worldbank.org/indicator/SP.POP.TOTL?locations=NG-US.

8. United Nations Department of Economic and Social Affairs, *World Population Prospects, the 2015 Revision* (2015).

9. BP, *Statistical Review of World Energy 2019* (British Petroleum, 2019); IEA, *World Energy Outlook 2019* (IEA, 2019); A. Kahan, "EIA Projects Nearly 50% Increase in World Energy Usage by 2050, Led by Growth in Asia" (US Energy Information Administration, 2019), https://www.eia.gov/todayinenergy/detail .php?id=41433#; V. Smil, *Energy Transitions: History, Requirements, Prospects* (Westport, CT: Praeger, 2010); H. Ritchie, *Our World in Data* (2014), https:// ourworldindata.org/energy.

10. BP, *Statistical Review of World Energy 2020* (British Petroleum, 2020).

11. BP, *Statistical Review of World Energy 2020* (British Petroleum, 2020).

12. D. J. C. MacKay, *Sustainable Energy—Without the Hot Air* (UIT Cambridge, 2009).

13. BP, *Statistical Review of World Energy 2020* (British Petroleum, 2020); D. J. C. MacKay, *Sustainable Energy—Without the Hot Air* (UIT Cambridge, 2009).

14. J. Heissel, C. Persico, and D. Simon, *Does Pollution Drive Achievement? The Effect of Traffic Pollution on Academic Performance* (National Bureau of Economic Research), http://www.nber.org/papers/w25489 (2019) doi:10.3386/w25489; J. G. Ayres and J. F. Hurley, *The Mortality Effects of Long-Term Exposure to Particulate Air Pollution in the United Kingdom* (UK Department for Environment, Food & Rural Affairs: Committee on the Medical Effects of Air Pollutants, 2010); European Environment Agency, *Air Pollution Fact Sheet 2013—United Kingdom* (European Union, 2013); "Air Pollution Deaths Are Double Previous Estimates, Finds Research," *Guardian* (2019), https://www.theguardian.com /environment/2019/mar/12/air-pollution-deaths-are-double-previous-esti mates-finds-research.

15. J. Cook et al., "Quantifying the Consensus on Anthropogenic Global Warming in the Scientific Literature," *Environmental Research Letters* 8 (2013): 024024; T. Stocker et al., *Climate Change 2013: The Physical Science Basis—Summary for Policymakers* (Intergovernmental Panel on Climate Change, 2013); D. M. Etheridge et al., "Natural and Anthropogenic Changes in Atmospheric CO_2 over the Last 1000 Years from Air in Antarctic Ice and Firn," *Journal of Geophysical Research: Atmospheres* 101 (1996): 4115–128; NOAA ESRL Global Monitoring Division, *Atmospheric Carbon Dioxide Dry Air Mole Fractions from Quasi-Continuous Measurements at Mauna Loa, Hawaii* (2014).

16. H. E. Huppert and R. S. J. Sparks, "Extreme Natural Hazards: Population Growth, Globalization and Environmental Change," *Philosophical Transactions of the Royal Society of London A: Mathematical, Physical and Engineering Sciences* 364 (2006): 1875–888.

17. Berkeley Earth, *Global Temperature Report for 2019* (Berkeley Earth, 2020), http://berkeleyearth.org/archive/2019-temperatures/; Z. Hausfather, *State of the Climate: How the World Warmed in 2019* (Carbon Brief, 2020), https://www .carbonbrief.org/state-of-the-climate-how-the-world-warmed-in-2019; World Health Organization, *Global Health Risks* (World Health Organization, 2009).

18. "Climate Change: Where We Are in Seven Charts and What You Can Do to Help," BBC News (2019), https://www.bbc.co.uk/news/science-environment -46384067; V. P. Masson-Delmotte et al., "Summary for Policymakers," in

Global Warming of 1.5C. An IPCC Special Report on the Impacts of Global Warming of 1.5C Above Pre-Industrial Levels and Related Global Greenhouse Gas Emission Pathways, in the Context of Strengthening the Global Response to the Threat of Climate Change, Sustainable Development, and Efforts to Eradicate Poverty (IPCC; World Meteorological Organization, 2018).

19. BP, *Statistical Review of World Energy 2020* (British Petroleum, 2020).

20. BP, *Statistical Review of World Energy 2020* (British Petroleum, 2020); V. Smil, *Energy Transitions: History, Requirements, Prospects* (Westport, CT: Praeger, 2010); H. Ritchie, *Our World in Data* (2014), https://ourworldindata.org/energy.

21. R. Fouquet, *Heat, Power and Light: Revolutions in Energy Services* (Cheltenham, UK: Edward Elgar Publishing Limited, 2008); V. Smil, *Energy Transitions: History, Requirements, Prospects* (Westport, CT: Praeger, 2010); H. Ritchie, "Energy," *Our World in Data* (2014), https://ourworldindata.org/energy; BP, *Statistical Review of World Energy 2020* (British Petroleum, 2020); International Energy Agency, *Key World Energy Statistics* (International Energy Agency, 2014); "The Pandas Development Team," *Pandas-dev/pandas: Pandas*, Zenodo, 2020, doi:10.5281/zenodo.3509134; J. D. Hunter, "Matplotlib: A 2D Graphics Environment," *Computing in Science & Engineering* 9 (2007): 90–95.

22. R. Fouquet and P. J. Pearson, "Seven Centuries of Energy Services: The Price and Use of Light in the United Kingdom (1300–2000)," *Energy Journal* 139 (2006): 177; I. MacLeay, K. Harris, and A. Annut, *Digest of United Kingdom Energy Statistics 2013* (UK Department of Energy; Climate Change, 2013); A. Kharina and D. Rutherford, *Fuel Efficiency Trends for New Commercial Jet Aircraft: 1960 to 2014* (The International Council on Clean Transportation, 2015).

23. L. A. Greening, D. L. Greene, and C. Difiglio, "Energy Efficiency and Consumption—The Rebound Effect: A Survey," *Energy Policy* 28 (2000): 389–401; H. Herring, "Is Energy Efficiency Environmentally Friendly?," *Energy & Environment* 11 (2000): 313–325; S. B. Bruns, A. Moneta, and D. Stern, *Macroeconomic Time-Series Evidence That Energy Efficiency Improvements Do Not Save Energy* (Centre for Applied Macroeconomic Analysis, Crawford School of Public Policy, The Australian National University, 2019), https://EconPapers.repec.org/RePEc:een:camaaa:2019-21.

24. J. J. Andersson, "Carbon Taxes and CO_2 Emissions: Sweden as a Case Study," *American Economic Journal: Economic Policy* 11 (2019): 1–30; A. Yamazaki, "Jobs and Climate Policy: Evidence from British Columbia's Revenue-Neutral Carbon Tax," *Journal of Environmental Economics and Management* 83 (2017): 197–216.

25. *Initiative on Global Markets. Surveys of Economists on Carbon Taxes* (University of Chicago Booth School of Business, 2020), https://www.igmchicago.org/?s=carbon+tax; P. H. Howard and D. Sylvan, "The Economic Climate: Establishing Expert Consensus on the Economics of Climate Change," *Institute for Policy Integrity* (2015): 438–41; N. G. Mankiw, "Smart Taxes: An Open Invitation to Join the Pigou Club," *Eastern Economic Journal* 35 (2009): 14–23.

26. J. Rogleji et al., "2018: Mitigation Pathways Compatible with 1.5C in the Context of Sustainable Development," in *Global Warming of 1.5C. An IPCC Special Report on the Impacts of Global Warming of 1.5C Above Pre-Industrial Levels and Related Global Greenhouse Gas Emission Pathways, in the Context of*

Strengthening the Global Response to the Threat of Climate Change, Sustainable Development, and Efforts to Eradicate Poverty (IPCC, 2018).

27. International Renewable Energy Agency, *How Falling Costs Make Renewables a Cost-Effective Investment* (International Renewable Energy Agency, 2020), https://www.irena.org/newsroom/articles/2020/Jun/How-Falling-Costs-Make -Renewables-a-Cost-effective-Investment; UK Government, *Electricity Generation Costs 2020* (Department of Business, Energy and Industrial Strategy, 2020).

28. L. M. Miller and D. W. Keith, "Corrigendum: Observation-Based Solar and Wind Power Capacity Factors and Power Densities," *Environmental Research Letters* 14 (2019): 079501.

29. D. J. C. MacKay, *Sustainable Energy—Without the Hot Air* (UIT Cambridge, 2009); "The Great Myth of Urban Britain," BBC News (2012), http://www.bbc .co.uk/news/uk-18623096.

30. M. Dröes and H. R. A. Koster, *Wind Turbines, Solar Farms, and House Prices* (C.E.P.R. Discussion Papers, 2020), https://EconPapers.repec.org /RePEc:cpr:ceprdp:15023; G. Meyer, "The US and Climate: New York's Bold Green Plans Hit Opposition," *Financial Times* (2020), https://www.ft.com /content/61a07f4f-1622-4bea-a71d-f927cf113636.

31. J. Gummer et al., *Net Zero—Technical Report* (UK Committee on Climate Change, 2019).

32. G. Myhre et al., "Frequency of Extreme Precipitation Increases Extensively with Event Rareness Under Global Warming," *Scientific Reports* 9 (2019): 1–10; K. Solaun and E. Cerdá, "Climate Change Impacts on Renewable Energy Generation. A Review of Quantitative Projections," *Renewable and Sustainable Energy Reviews* 116 (2019): 109415.

33. P. Denholm, M. O'Connell, G. Brinkman, and J. Jorgenson, *Overgeneration from Solar Energy in California. A Field Guide to the Duck Chart* (National Renewable Energy Lab [NREL], Golden, CO, 2015).

34. O. Edenhofer, R. Pichs-Madruga, Y. Sokona, and K. Seyboth, "Summary for Policymakers," in *Special Report on Renewable Energy Sources and Climate Change Mitigation of Global Warming of 1.5C Above Pre-Industrial Levels and Related Global Greenhouse Gas Emission Pathways, in the Context of Strengthening the Global Response to the Threat of Climate Change, Sustainable Development, and Efforts to Eradicate Poverty* (IPCC, 2011); P. Moriarty and D. Honnery, "What Is the Global Potential for Renewable Energy?," *Renewable and Sustainable Energy Reviews* 16 (2012): 244–52; P. Moriarty and D. Honnery, "Can Renewable Energy Power the Future?," *Energy Policy* 93 (2016): 3–7; J. D. Jenkins, M. Luke, and S. Thernstrom, "Getting to Zero Carbon Emissions in the Electric Power Sector," *Joule* 2 (2018): 2498–510.

35. M. Pehl et al. "Understanding Future Emissions from Low-Carbon Power Systems by Integration of Life-Cycle Assessment and Integrated Energy Modelling," *Nature Energy* 2 (2017): 939.

36. Ipsos-Mori, *Global Citizen Reaction to the Fukushima Nuclear Plant Disaster* (Ipsos-Mori, 2011).

37. "UK Renewable Energy Auction Prices Plunge," *Financial Times* (2019), https:// www.ft.com/content/472e18cc-db7a-11e9-8f9b-77216ebe1f17.

38. M. Pehl et al., "Understanding Future Emissions from Low-Carbon Power Sys-

tems by Integration of Life-Cycle Assessment and Integrated Energy Modelling," *Nature Energy* 2 (2017): 939.

39. BP, *Statistical Review of World Energy 2019* (British Petroleum, 2019).

40. U. Bardi, "Extracting Minerals from Seawater: An Energy Analysis," *Sustainability* 2 (2010): 980–92.

41. BP, *Statistical Review of World Energy 2020* (British Petroleum, 2020); Pandas Development Team, *Pandas-dev/Pandas: Pandas*, Zenodo, 2020, doi:10.5281/zenodo.3509134; J. D. Hunter, "Matplotlib: A 2D Graphics Environment," *Computing in Science & Engineering* 9 (2007): 90–95; S. Fetter, "How Long Will the World's Uranium Supplies Last?," *Scientific American* (2009), https://www.scientificamerican.com/article/how-long-will-global-uranium-deposits-last/; Nuclear Energy Agency and the International Atomic Energy Agency, *Uranium 2018: Resources, Production and Demand* (OECD, 2019), https://doi.org/10.1787/uranium-2018-en; A. M. Bradshaw, T. Hamacher, and U. Fischer, "Is Nuclear Fusion a Sustainable Energy Form?," *Fusion Engineering and Design* 86 (2011): 2770–773.

42. C. Liu et al., "Lithium Extraction from Seawater Through Pulsed Electrochemical Intercalation," *Joule* 4 (2020): 1459–469.

43. K. Bourzac, "Fusion Start-ups Hope to Revolutionize Energy in the Coming Decades," *Chemical Engineering News* (2018), https://cen.acs.org/energy/nuclear-power/Fusion-start-ups-hope-revolutionize/96/i32.

44. J. A. Etzler, *The Paradise Within the Reach of All Men: Without Labor, by Powers of Nature and Machinery* (J. Cleave, 1842); "SOLAR Energy: What the Sun's Rays Can Do and May Yet Be Made to Do," *Washington Star* (1891); "Use of Solar Energy Is Near a Solution; German Scientist's Improved Device Held to Rival Hydroelectric," *New York Times* (1931).

Chapter 3: Energy from Atoms

1. A. S. Eddington, "The Internal Constitution of the Stars," *Observatory* 43 (1920): 341–58.

2. M. Poole, J. Dainton, and S. Chattopadhyay, "Cockcroft's Subatomic Legacy: Splitting the Atom," *CERN Courier* (2007), https://cerncourier.com/a/cockcrofts-subatomic-legacy-splitting-the-atom/; N. Bohr, "The Rutherford Memorial Lecture 1958: Reminiscences of the Founder of Nuclear Science and of Some Developments Based on His Work," *Proceedings of the Physical Society* 78 (1961): 1083–115; R. H. Cragg, "Lord Ernest Rutherford of Nelson (1871–1937)," *Royal Institute of Chemistry Reviews* 4 (1971): 129–45; M. Kumar, "The Man Who Went Nuclear: How Ernest Rutherford Ushered in the Atomic Age," *Independent* (2011), https://www.independent.co.uk/news/science/the-man-who-went-nuclear-how-ernest-rutherford-ushered-in-the-atomic-age-2230533.html; J. K. Laylin, *Nobel Laureates in Chemistry, 1901–1992* (Chemical Heritage Foundation, 1993); H. R. Robinson, "Rutherford: Life and Work to the Year 1919, with Personal Reminiscences of the Manchester Period," *Proceedings of the Physical Society* 55 (1943): 161–82; M. A. Ainslie,

Principles of Sonar Performance Modelling (Springer, 2010); C. Jarlskog, "Lord Rutherford of Nelson, His 1908 Nobel Prize in Chemistry, and Why He Didn't Get a Second Prize," *Journal of Physics: Conference Series* 136 (2008): 012001.

3. J. Navarro, *A History of the Electron: JJ and GP Thomson* (Cambridge University Press, 2012).

4. E. Rutherford, "The Scattering of α and β Particles by Matter and the Structure of the Atom," *Philosophical Magazine* 92 (1911): 379–98; E. N. da C. Andrade, *Rutherford and the Nature of the Atom*, vol. 35 (Gloucester, MA: Peter Smith Publisher, 1964); W. E. Burcham, "Nuclear Physics in the United Kingdom 1911–1986," *Reports on Progress in Physics* 52 (1989): 823–79; E. Rutherford and H. Geiger, "The Charge and Nature of the α-particle," *Proceedings of the Royal Society of London. Series A, Containing Papers of a Mathematical and Physical Character* 81 (1908): 162–73.

5. B. Bryson, *A Short History of Nearly Everything* (Kottayam, India: DC Books, 2003).

6. J. Blackmore, "Ernst Mach Leaves 'the Church of Physics,'" *British Journal for the Philosophy of Science* 40 (1989): 519–540.

7. E. Rutherford, "LIV Collision of α Particles with Light Atoms IV. An anomalous Effect in Nitrogen," *London, Edinburgh, and Dublin Philosophical Magazine and Journal of Science* 37 (1919): 581–87; P. M. S. Blackett, "The Ejection of Protons from Nitrogen Nuclei, Photographed by the Wilson Method," *Proceedings of the Royal Society of London. Series A, Containing Papers of a Mathematical and Physical Character* 107 (1925): 349–60.

8. M. Poole, J. Dainton, and S. Chattopadhyay, "Cockcroft's Subatomic Legacy: Splitting the Atom," *CERN Courier* (2007), https://cerncourier.com/a/cockcrofts-subatomic-legacy-splitting-the-atom/; J. Cockcroft, and E. Walton, "Experiments with High Velocity Positive Ions ii. The Disintegration of Elements by High Velocity Protons," *Proceedings of the Royal Society of London. Series A, Mathematical, Physical and Engineering Sciences* 137 (1932): 229–42; J. Cockcroft and E. Walton, "Experiments with High Velocity Positive Ions," *Proceedings of the Royal Society of London. Series A, Containing Papers of a Mathematical and Physical Character* 129 (1930): 477–89.

9. R. Herman, *Fusion: The Search for Endless Energy* (Cambridge University Press, 1990); M. L. E. Oliphant, P. Harteck, and E. Rutherford, "Transmutation Effects Observed with Heavy Hydrogen," *Proceedings of the Royal Society of London. Series A, Containing Papers of a Mathematical and Physical Character* 144 (1934): 692–703.

10. R. Sherr, K. T. Bainbridge, and H. H. Anderson, "Transmutation of Mercury by Fast Neutrons," *Physical Review* 60 (1941): 473–79.

11. A. Einstein, "Does the Inertia of a Body Depend on Its Energy Content?," *Annalen der Physik* 323 (1905): 639–41; F. W. Dyson, A. S. Eddington, and C. Davidson, "A Determination of the Deflection of Light by the Sun's Gravitational Field, from Observations Made at the Total Eclipse of May 29, 1919," *Philosophical Transactions of the Royal Society of London. Series A: Mathematical, Physical and Engineering Sciences* 220 (1920): 291–333.

12. J. Cockcroft and E. Walton, "Experiments with High Velocity Positive Ions.

NOTES

ii. The Disintegration of Elements by High Velocity Protons," *Proceedings of the Royal Society of London. Series A, Mathematical, Physical and Engineering Sciences* 137 (1932): 229–242; M. L. E. Oliphant, P. Harteck, and E. Rutherford, "Transmutation Effects Observed with Heavy Hydrogen," *Proceedings of the Royal Society of London. Series A, Containing Papers of a Mathematical and Physical Character* 144 (1934): 692–703.

13. "Einstein's Equation of Life and Death," *BBC Horizon*, BBC (2014), http://www.bbc.co.uk/sn/tvradio/programmes/horizon/einstein_equation_trans.shtml; W. Kaempffert, "Rutherford Cools Atom Energy Hope," *New York Times*, 1933.

14. L. Spitzer, Jr., *Physics of Fully Ionised Gases* (Geneva: Interscience, 1967).

15. H. Alfvén, *Nobel Lectures: Physics 1963–1970* (Amsterdam: Elsevier Publishing Company, 1972).

16. J. Nuckolls, "Contributions to the Genesis and Progress of ICF," in *Inertial Confinement Nuclear Fusion: A Historical Approach by Its Pioneers* (eds. G. Velarde and N. Carpintero-Santamaria) (London: Foxwell & Davies, 2007); D. Clery, *A Piece of the Sun: The Quest for Fusion Energy* (New York: Abrams, 2014).

Chapter 4: How the Universe Builds Stars

1. A. Eddington, *Stars and Atoms* (Oxford: Clarendon Press, 1927).

2. H. Johnston, Lives of the Stars Lectures: Star Birth (2016).

3. L. Koopmans et al., "The Cosmic Dawn and Epoch of Reionization with the Square Kilometre Array," *arXiv preprint arXiv:1505.07568* (2015); A. Patil et al., "Upper Limits on the 21 cm Epoch of Reionization Power Spectrum from One Night with LOFAR," *Astrophysical Journal* 838 (2017): 65.

4. E. M. Burbidge, G. R. Burbidge, W. A. Fowler, and F. Hoyle, "Synthesis of the Elements in Stars," *Reviews of Modern Physics* 29 (1957): 547.

5. O. Benomar et al., "Asteroseismic Detection of Latitudinal Differential Rotation in 13 Sun-like Stars," *Science* 361 (2018): 1231–234.

6. "The Hidden Mechanics of Magnetic Field Reconnection, A Key Factor in Solar Storms and Fusion Energy Reactors," Phys.org (2017), https://phys.org/news/2017-10-hidden-mechanics-magnetic-field-reconnection.html; NASA, "The Day the Sun Brought Darkness" (2009), https://www.nasa.gov/topics/earth/features/sun_darkness.html.

7. F. J. Dyson, "Search for Artificial Stellar Sources of Infrared Radiation" *Science* 131 (1960): 1667–668.

8. H. Johnston, "Lives of the Stars Lectures: Star Birth" (2016); F. Tramper et al., "Massive Stars on the Verge of Exploding: The Properties of Oxygen Sequence Wolf-Rayet Stars," *Astronomy & Astrophysics* 581 (2015): A110.

9. A. Eddington, *Stars and Atoms* (Oxford: Clarendon Press, 1927).

10. K. P. Schröder and R. Connon Smith, "Distant Future of the Sun and Earth Revisited," *Monthly Notices of the Royal Astronomical Society* 386 (2008): 155–63.

11. P. F. Winkler, G. Gupta, and K. S. Long, "The SN 1006 Remnant: Optical Proper Motions, Deep Imaging, Distance, and Brightness at Maximum," *Astrophysical Journal* 585 (2003): 324; N. Gehrels et al., "Ozone Depletion from Nearby Su-

pernovae," *Astrophysical Journal* 585 (2003): 1169; B. R. Goldstein, "Evidence for a Supernova of AD 1006," *Astronomical Journal* 70 (1965): 105; W. Rada and R. Neuhaeuser, "Supernova SN 1006 in Two Historic Yemeni Reports," *Astronomische Nachrichten* 336 (2015): 249–57.

Chapter 5: How to Build a Star with Magnetic Fields

1. D. J. C. MacKay, *Sustainable Energy—Without the Hot Air* (UIT Cambridge, 2009); M. Kaku, *Physics of the Impossible: A Scientific Exploration into the World of Phasers, Force Fields, Teleportation, and Time Travel* (New York: Anchor, 2009), 46–47.

2. *Lords Sitting—JET Nuclear Fusion Project: HL deb.*, vol. 485, 1517–1519 (Houses of Parliament, 1987).

3. J. J. Thomson, *The Corpuscular Theory of Matter* (London: A. Constable & Company, Limited, 1907).

4. X. Litaudon et al., "Overview of the JET Results in Support to ITER," *Nuclear Fusion* 57 (2017): 102001.

5. P. H. Rebut, "The Joint European Torus (JET)," *European Physical Journal H* 43 (2018): 459–97.

6. M. Steenbeck and K. Hoffmann, *Siemens Technical Report HW/PL*: Number 27 (1943).

7. W. H. Bennett, "Magnetically Self-Focussing Streams," *Physical Review* 45 (1934): 890.

8. A. Ware, "A Study of High-Current Toroidal Ring Discharge," *Philosophical Transactions of the Royal Society of London A: Mathematical, Physical and Engineering Sciences* 243 (1951): 197–220; S. Cousins and A. Ware, "Pinch Effect Oscillations in a High Current Toroidal Ring Discharge," *Proceedings of the Physical Society B* 64 (1951): 159; R. Carruthers, "The Beginning of Fusion at Harwell," *Plasma Physics and Controlled Fusion* 30 (1988): 1993; T. Török, B. Kliem, and V. Titov, "Ideal Kink Instability of a Magnetic Loop Equilibrium," *Astronomy & Astrophysics* 413 (2004): L27–L30.

9. J. D. Hunter, "Matplotlib: A 2D Graphics Environment," *Computing in Science & Engineering* 9 (2007): 90–95.

10. J. Wesson and D.J. Campbell, *Tokamaks*, vol. 149 (Oxford: Oxford University Press, 2011).

11. F. Chen, *An Indispensable Truth: How Fusion Power Can Save the Planet* (London: Springer Science + Business Media, 2011).

12. H. Weisen et al., "The Scientific Case for a JET DT Experiment," in *AIP Conference Proceedings*, vol. 1612 (AIP, 2014), 77–86.

13. J. D. Lawson, "Some Criteria for a Power Producing Thermonuclear Reactor," *Proceedings of the Physical Society B* 70 (1957): 6.

Chapter 6: How to Build a Star with Inertia

1. R. Herman, *Fusion: The Search for Endless Energy* (Cambridge University Press, 1990).
2. W. Dunn, "The New Industry of Building Stars," *New Statesman* (2018), https://www.newstatesman.com/spotlight/energy/2018/11/new-industry-building-stars.
3. E. Teller, *Energy from Heaven and Earth* (New York: W. H. Freeman Co. Limited, 1979).
4. E. Teller, "Comments on Plasma Stability and on a Constant-Pressure Thermonuclear Reactor," in *Conference on Thermonuclear Reactions, Princeton University, October 26–27, 1954*; A. S. Bishop, *Project Sherwood: The US Program in Controlled Fusion* (Boston: Addison-Wesley, 1958).
5. J. Nuckolls, "Contributions to the Genesis and Progress of ICF," in *Inertial Confinement Nuclear Fusion: A Historical Approach by Its Pioneers* (eds. G. Velarde and N. Carpintero-Santamaria) (London: Foxwell & Davies, 2007).
6. T. H. Maiman, "Stimulated Optical Radiation in Ruby," *Nature* 187 (1960): 493.
7. J. Nuckolls, "Contributions to the Genesis and Progress of ICF," in *Inertial Confinement Nuclear Fusion: A Historical Approach by Its Pioneers* (eds. G. Velarde and N. Carpintero-Santamaria) (London: Foxwell & Davies, 2007); D. Clery, *A Piece of the Sun: The Quest for Fusion Energy* (New York: Abrams, 2014).
8. R. Kidder, "Laser Fusion: The First Ten Years 1962–1972," in *Inertial Confinement Nuclear Fusion: A Historical Approach by Its Pioneers* (eds. G. Velarde and N. Carpintero-Santamaria) (London: Foxwell & Davies, 2007).
9. J. D. Lindl, "Development of the Indirect-Drive Approach to Inertial Confinement Fusion and the Target Physics Basis for Ignition and Gain," *Physics of Plasmas* 2 (1995): 3933–4024; R. Craxton et al., "Direct-Drive Inertial Confinement Fusion: A Review," *Physics of Plasmas* 22 (2015): 110501.
10. S. Atezni and J. Meyer-ter-Vehn, *The Physics of Inertial Fusion* (Oxford Science Publications, 2004); J. D. Lindl, "Development of the Indirect-Drive Approach to Inertial Confinement Fusion and the Target Physics Basis for Ignition and Gain," *Physics of Plasmas* 2 (1995): 3933–4024; R. Olson et al., "First Liquid Layer Inertial Confinement Fusion Implosions at the National Ignition Facility," *Physical Review Letters* 117 (2016): 245001.
11. J. D. Lindl, "Development of the Indirect-Drive Approach to Inertial Confinement Fusion and the Target Physics Basis for Ignition and Gain," *Physics of Plasmas* 2 (1995): 3933–4024.
12. H. Robey et al., "The Effect of Laser Pulse Shape Variations on the Adiabat of NIF Capsule Implosions," *Physics of Plasmas* 20 (2013): 052707.
13. F. Suzuki-Vidal et al., "Interaction of a Supersonic, Radiatively Cooled Plasma Jet with an Ambient Medium," *Physics of Plasmas* 19 (2012): 022708.
14. D. H. Sharp, *Overview of Rayleigh-Taylor Instability* (Los Alamos National Lab., NM, USA, 1983).
15. R. Herman, *Fusion: The Search for Endless Energy* (Cambridge University Press, 1990); J. D. Lindl, "Development of the Indirect-Drive Approach to Inertial

Confinement Fusion and the Target Physics Basis for Ignition and Gain," *Physics of Plasmas* 2 (1995): 3933–4024; C. M. Braams and P. E. Stott, *Nuclear Fusion: Half a Century of Magnetic Confinement Fusion Research* (Bristol, UK: Institute of Physics Publishing, 2002); G. McCracken and P. Stott, *Fusion: The Energy of the Universe* (Amsterdam: Elsevier, 2005); R. G. Evans, "UK Fusion Breakthrough Revealed at Last," *Physics World* 23 (2010); C. Marsh, P. D. Roberts, and K. Johnston, "Nuclear Testing: A UK Perspective," in *US-UK Nuclear Cooperation After 50 Years* (Washington, DC: CSIS Press, 2008), 228; W. J. Broad, "Secret Advance in Nuclear Fusion Spurs a Dispute Among Scientists," *New York Times* (1988); G. Velarde and N. Carpintero-Santamaria, *Inertial Confinement Fusion: A Historical Approach by Its Pioneers* (London: Foxwell & Davies, 2007); S. Rose and J. Wark, "Laser Fusion's Hot Secrets Revealed," *Physics World* 7 (1994): 26.

Chapter 7: The New Star Builders

1. Quotations from Werner von Siemens, 1854–1892 (selection), Siemens Historical Institute (2016), https://assets.new.siemens.com/siemens/assets/api/uuid:979a30ef7b0cf73bdf3fa4eefed7cbabcbef8a10/070-werner-von-siemens-quotations1.pdf.
2. A. Riley, "This Shrimp Is Carrying a Real-Life Working Stun Gun," BBC (2016), http://www.bbc.co.uk/earth/story/20160129-the-shrimp-that-has-turned-bubbles-into-a-lethal-weapon; M. Webster, "Bigger Bacon," *Radiolab from WNYC* (2016), https://www.wnycstudios.org/podcasts/radiolab/articles/bigger-bacon; M. Versluis, B. Schmitz, A. von der Heydt, and D. Lohse, "How Snapping Shrimp Snap: Through Cavitating Bubbles," *Science* 289 (2000): 2114–117.
3. C. MacLeod, "The 1690s Patents Boom: Invention or Stock-Jobbing?," *Economic History Review* (1986): 549–71.
4. "Oxford Startup Promises Fusion Gain by 2024," *EENews Europe* (2019), https://www.eenewseurope.com/news/oxford-startup-promises-fusion-gain-2024#.
5. A. Sykes et al., "Compact Fusion Energy Based on the Spherical Tokamak," *Nuclear Fusion* 58 (2017): 016039; A. Sykes, "Progress on Spherical Tokamaks," *Plasma Physics and Controlled Fusion* 36 (1994): B93.
6. T. Lee, I. Jenkins, E. Surrey, and D. Hampshire, "Optimal Design of a Toroidal Field Magnet System and Cost of Electricity Implications for a Tokamak Using High Temperature Superconductors," *Fusion Engineering and Design* 98 (2015): 1072–75; D. Cohn and L. Bromberg, "Advantages of High-Field Tokamaks for Fusion Reactor Development," *Journal of Fusion Energy* 5 (1986): 161–70.
7. A. Creely et al., "Overview of the SPARC Tokamak," *Journal of Plasma Physics* 86 (2020).
8. "Nuclear Fusion on Brink of Being Realised, Say MIT Scientists," *Guardian* (2018), https://www.theguardian.com/environment/2018/mar/09/nuclear-fusion-on-brink-of-being-realised-say-mit-scientists; B. N. Sorbom et al., "ARC: A Compact, High-Field, Fusion Nuclear Science Facility and Demonstration Power Plant with Demountable Magnets," *Fusion Engineering and Design* 100 (2015): 378–405.

NOTES

9. S. Wurzel, "The Number of Fusion Energy Startups Is Growing Fast—Here's Why," Fusion Energy Base (2020), https://www.fusionenergybase.com/article /the-number-of-fusion-energy-startups-is-growing-fast-heres-why/; D. Chandler, "MIT and Newly Formed Company Launch Novel Approach to Fusion Power," *MIT News* (2018), https://news.mit.edu/2018/mit-newly -formed-company-launch-novel-approach-fusion-power-0309; J. Tollefson, "MIT Launches Multimillion-Dollar Collaboration to Develop Fusion Energy," *Nature* 555 (2018); J. Shieber, "A Boston Startup Developing a Nuclear Fusion Reactor Just Got a Roughly $50 Million Boost," Tech Crunch (2019), https://techcrunch.com/2019/06/27/a-boston-startup-developing-a-nuclear -fusion-reactor-just-got-a-roughly-50-million-boost.

10. "The British Reality TV Star Building a Fusion Reactor," BBC News (2017), https://www.bbc.com/future/article/20170418-the-made-in-chelsea-star -building-a-fusion-reactor.

11. ARPA-E, "Department of Energy Announces $32 million for Lower-Cost Fusion Concepts" (2020), https://arpa-e.energy.gov/news-and-media/press -releases/department-energy-announces-32-million-lower-cost-fusion -concepts; ARPA-E, "Department of Energy Announces $29 million in Fusion Energy Technology Development" (2020), https://arpa-e.energy.gov/news -and-media/press-releases/department-energy-announces-29-million-fusion -energy-technology.

12. K. Graham, "Canadian Startup Gets a $65 Million Boost for Fusion Power Plant," *Digital Journal* (2019), http://www.digitaljournal.com/tech-and-science /technology/canadian-startup-gets-a-65-million-boost-for-fusion-power -plant/article/563880; T. Orton, "General Fusion Partners with Hatch for Prototype Power Plant," JWN Energy (2020), https://www.jwnenergy.com /article/2020/1/general-fusion-partners-hatch-prototype-power-plant/.

13. M. Delage, "Timing Is Everything: Pushing Fusion Forward with Pistons & Cutting-Edge Electronics," General Fusion (2018), https://generalfusion .com/2018/11/timing-everything-pushing-fusion-forward-pistons-cutting -edge-electronics/; M. Laberge, "Magnetized Target Fusion with a Spherical Tokamak," *Journal of Fusion Energy* 38 (2019): 199–203; B. Borzykowski, "Why Bezos and Microsoft Are Betting on This $10 Trillion Energy Fix for the Planet," CNBC (2019), https://www.cnbc.com/2019/03/06/bezos-microsoft-bet-on-a-10-trillion -energy-fix-for-the-planet.html; T. Hamilton, "A New Approach to Fusion," *MIT Technology Review* (2009), https://www.technologyreview.com/s/414559/a-new -approach-to-fusion/; D. Robitzski, "Expert: 'I'm 100 Percent Confident' Fusion Power Will Be Practical," Futurism (2019), https://futurism.com/fusion-power -practical; T. Orton, "General Fusion Raises $27m, Construction on Large-Scale Prototype Two Years Away," *Business Vancouver* (2015), https://biv.com /article/2015/05/general-fusions-raises-27m-ceo-says-prototype-two-.

14. J. McMahon, "Energy from Fusion in a Couple Years, CEO Says, Commercialization in Five," *Forbes* (2019), https://www.forbes.com/sites /jeffmcmahon/2019/01/14/private-firm-will-bring-fusion-reactor-to-market -within-five-years-ceo-says/#56c240f11d4a; M. Kanellos, "Hollywood, Silicon Valley and Russia Join Forces on Nuclear Fusion," *Forbes* (2013), https://www

.forbes.com/sites/michaelkanellos/2013/03/11/hollywood-silicon-valley-and
-russia-join-forces-on-nuclear-fusion/#4ff137e272ba; B. Wang, "CEO of TAE
Technologies Says They Will Begin Commercialization of Fusion by 2023,"
NextBigFuture (2019), https://www.nextbigfuture.com/2019/01/ceo-of-tae
-technologies-says-they-will-reach-commercial-fusion-by-2024.html.

15. M. Haines, "Plasma Containment in Cusp-Shaped Magnetic Fields," *Nuclear Fusion* 17 (1977): 811; S. Best, "Trouble for Lockheed's Fusion Reactor?," *Daily Mail* (2017), https://www.dailymail.co.uk/sciencetech/article-4473908/Trouble-Lockheed-s-fusion-reactor.html; J. Trevithik, "Skunk Works' Exotic Fusion Reactor Program Moves Forward with Larger, More Powerful Design," *The Drive* (2019), https://www.thedrive.com/the-war-zone/29074/skunk-works-exotic-fusion-reactor-program-moves-forward-with-larger-more-powerful-design; R. Smith, "Lockheed Martin Doubles Down on Cold Fusion," Yahoo! Finance (2019), https://finance.yahoo.com/news/lockheed-martin-doubles-down-cold-120300203.html.

16. "Startup Nuclear Energy Companies Augur Safer, Cheaper Atomic Power," *Fortune* (2014), https://fortune.com/2014/07/03/startup-nuclear-energy-companies/; "A New Approach to Fusion," *MIT Technology Review* (2009), https://www.technologyreview.com/s/414559/a-new-approach-to-fusion/; BIV, "General Fusion Raises 27m, Construction on Large-Scale Prototype Two Years Away," BIV.com (2015), https://biv.com/article/2015/05/general-fusions-raises-27m-ceo-says-prototype-two-; Futurism, "Expert: 'I'm 100 Percent Confident' Fusion Power Will Be Practical," Futurism.com (2019), https://futurism.com/fusion-power-practical.

17. V. J. Stenger, "Is the Big Bang a Bust?," *Skeptical Inquirier* 16 (1992); A. Penzias, "Big Bang Theory Makes Sense of Cosmic Facts; No Contradiction," *New York Times* (1991), https://www.nytimes.com/1991/06/18/opinion/l-big-bang-theory-makes-sense-of-cosmic-facts-no-contradiction-092291.html; B. Feuerbacher and R. Scranton, "Evidence for the Big Bang" (2006), http://www.talkorigins.org/faqs/astronomy/bigbang.html#lerner.

18. D. Klir et al., "Ion Acceleration Mechanism in Mega-Ampere Gas-Puff Z-Pinches," *New Journal of Physics* 20 (2018): 053064; M. Krishnan, "The Dense Plasma Focus: A Versatile Dense Pinch for Diverse Applications," *IEEE Transactions on Plasma Science* 40 (2012): 3189–221.

19. M. Halper, "Startup Nuclear Energy Companies Augur Safer, Cheaper Atomic Power," *Fortune* (2014), https://fortune.com/2014/07/03/startup-nuclear-energy-companies/; M. Anderson, "How Far Can Crowd-Funded Nuclear Fusion Go?," IEEE Spectrum (2014), https://spectrum.ieee.org/energywise/energy/nuclear/how-far-can-crowdfunded-nuclear-fusion-go; E. J. Lerner, "Invest in LPPFusion," Wefunder (2017), https://wefunder.com/lppfusion/about; E. Lerner, *The Big Bang Never Happened: A Startling Refutation of the Dominant Theory of the Origin of the Universe* (New York: Vintage, 2010); E. J. Lerner, S. M. Hassan, I. Karamitsos, and F. V. Roessel, "Confined Ion Energy >200 keV and Increased Fusion Yield in a DPF with Monolithic Tungsten Electrodes and Pre-Ionization," *Physics of Plasmas* 24 (2017): 102708; E. J. Lerner, S. K. Murali, D. Shannon, A. M. Blake, and F. V. Roessel, "Fusion Reactions from >150 keV Ions in a Dense Plasma Focus Plasmoid," *Physics of Plasmas* 19 (2012):

032704; E. J. Lerner, "Thank you all!," Wefunder (2020), https://wefunder
.com/updates/130741-thank-you-all.

20. T. S. Pedersen et al., "Confirmation of the Topology of the Wendelstein 7-X
Magnetic Field to Better Than 1:100,000," *Nature Communications* 7 (2016).

21. Max-Planck-Gesellschaft, "Angela Merkel Switches on Wendelstein 7-X Fusion
Device" (2016), https://www.mpg.de/9926419/wendelstein7x-start; T. S. Ped-
ersen et al., "Confirmation of the Topology of the Wendelstein 7-X Magnetic
Field to Better Than 1:100,000," *Nature Communications* 7 (2016).

22. A. Sykes et al., "First Physics Results from the MAST Mega-Amp Spherical
Tokamak," *Physics of Plasmas* 8 (2001): 2101–106.

23. "Boris Johnson Jokes About UK Being on the Verge of Nuclear Fusion," *New
Scientist* (2019), https://www.newscientist.com/article/2218570-boris-johnson
-jokes-about-uk-being-on-the-verge-of-nuclear-fusion/#ixzz66tYUwh6k;
P. Ball, "A Lightbulb Moment for Nuclear Fusion?," *Guardian* (2019), https://
www.theguardian.com/environment/2019/oct/27/nuclear-fusion-research
-power-generation-iter-jet-step-carbon-neutral-2050-boris-johnson; E. Gibney,
"UK Hatches Plan to Build World's First Fusion Power Plant," *Nature* (2019),
https://www.nature.com/articles/d41586-019-03039-9.

24. A. Harvey-Thompson et al., "Diagnosing and Mitigating Laser Preheat Induced Mix in MagLIF," *Physics of Plasmas* 25 (2018): 112705; S. A. Slutz et al.,
"Scaling Magnetized Liner Inertial Fusion on Z and Future Pulsed-Power Accelerators," *Physics of Plasmas* 23 (2016): 022702; S. A. Slutz and R. A. Vesey,
"High-Gain Magnetized Inertial Fusion," *Physical Review Letters* 108 (2012):
025003; S. A. Slutz et al., "Pulsed-Power-Driven Cylindrical Liner Implosions
of Laser Preheated Fuel Magnetized with an Axial Field," *Physics of Plasmas* 17
(2010): 056303; M. V. Berry and A. K. Geim, "Of Flying Frogs and Levitrons,"
European Journal of Physics 18 (1997): 307.

25. T. Peckinpaugh, M. O'Neill, and A. Johns, "U.S. House of Representatives
Demonstrates Support for Fusion Energy" (2020), https://www.globalpower
lawandpolicy.com/2020/09/u-s-house-of-representatives-demonstrates
-support-for-fusion-energy/.

Chapter 8: Isn't This All a Bit Dangerous?

1. R. Rhodes, *Dark Sun: The Making of the Hydrogen Bomb* (New York: Simon &
Schuster, 1995); R. G. Hewlett and F. Duncan, *Atomic Shield, 1947–1952*, vol. 2
(Pennsylvania State University Press, 1969).

2. L. Engel, "Twenty-three Fishermen and a Bomb: The Voyage of the Lucky
Dragon," *New York Times* (1958); B. Kendall, "The H-Bomb," *Cold War: Stories
from the Big Freeze*, BBC Radio 4 (2016).

3. C. Bernardini and L. Bonolis, *Enrico Fermi: His Work and Legacy* (London:
Springer Science & Business Media, 2013).

4. A. Robock, L. Oman, and G. L. Stenchikov, "Nuclear Winter Revisited with a
Modern Climate Model and Current Nuclear Arsenals: Still Catastrophic Consequences," *Journal of Geophysical Research: Atmospheres* 112 (2007); M. Roser

and M. Nagdy, "Nuclear Weapons," Our World in Data (2013), https://ourworld indata.org/nuclear-weapons.

5. A. Glaser and R. J. Goldston, "Proliferation Risks of Magnetic Fusion Energy: Clandestine Production, Covert Production and Breakout," *Nuclear Fusion* 52 (2012): 043004.

6. A. Glaser and R. J. Goldston, "Proliferation Risks of Magnetic Fusion Energy: Clandestine Production, Covert Production and Breakout," *Nuclear Fusion* 52 (2012): 043004.

7. M. Claessens, *ITER: The Giant Fusion Reactor: Bringing a Sun to Earth* (London: Springer Nature, 2019).

8. R. H. Cragg, "Lord Ernest Rutherford of Nelson (1871–1937)," *Royal Institute of Chemistry, Review* 4 (1971): 129–45; E. Rutherford and T. Royds, "XXI. The Nature of the α Particle from Radioactive Substances," *London, Edinburgh, and Dublin Philosophical Magazine and Journal of Science* 17 (1909): 281–86; E. Rutherford, "VIII. Uranium Radiation and the Electrical Conduction Produced by It," *London, Edinburgh, and Dublin Philosophical Magazine and Journal of Science* 47 (1899): 109–63.

9. R. Blandford, P. Simeon, and Y. Yuan, "Cosmic Ray Origins: An Introduction," *Nuclear Physics B—Proceedings Supplements* 256–257 (2014): 9–22.

10. P. De Marcillac, N. Coron, G. Dambier, J. Leblanc, and J. P. Moalic, "Experimental Detection of α-Particles from the Radioactive Decay of Natural Bismuth," *Nature* 422 (2003): 876–78.

11. W. Friedberg, K. Copeland, F. E. Duke, K. O'Brien III, and E. B. Darden Jr., "Radiation Exposure During Air Travel: Guidance Provided by the Federal Aviation Administration for Air Carrier Crews," *Health Physics* 79 (2000): 591–95.

12. Public Health England, *Ionising Radiation: Dose Comparisons* (UK Government, 2011), https://www.gov.uk/government/publications/ionising-radiation -dose-comparisons/ionising-radiation-dose-comparisons.

13. M. Hvistendahl, "Coal Ash Is More Radioactive Than Nuclear Waste," *Scientific American* 13 (2007); J. McBride, R. Moore, J. Witherspoon, and R. Blanco, *Radiological Impact of Airborne Effluents of Coal-Fired and Nuclear Power Plants* (Oak Ridge National Lab., Tenn., USA, 1977).

14. N. Jones, "Carbon Dating, the Archaeological Workhorse, Is Getting a Major Reboot," *Nature* (2020); S. S. Schweber and S. Schweber, *Nuclear Forces: The Making of the Physicist Hans Bethe* (Harvard University Press, 2012); J. W. Valley et al., "Hadean Age for a Post-Magma-Ocean Zircon Confirmed by Atom-Probe Tomography," *Nature Geoscience* 7 (2014): 219–23; T. Higham et al., "The Timing and Spatiotemporal Patterning of Neanderthal Disappearance," *Nature* 512 (2014): 306.

15. E. O. Lawrence, "Transmutations of Sodium by Deutons," *Physical Review* 47 (1935): 17.

16. OECD, *Physics and Safety of Transmutation Systems: A Status Report* (OECD, 2006).

17. L. N. Larson, *Nuclear Waste Storage Sites in the United States* (Congressional Research Service, 2020), https://fas.org/sgp/crs/nuke/IF11201.pdf; Nuclear Decommissioning Authority, "UK Radioactive Waste Inventory" (2020), https://ukinventory.nda.gov.uk/.

18. ITER Organisation, "Safety and Environment" (2020), https://www.iter.org/mach/safety.

19. M. García, P. Sauvan, R. García, F. Ogando, and J. Sanz, "Study of Concrete Activation with IFMIF-like Neutron Irradiation: Status of EAF and TENDL Neutron Activation Cross-Sections," in *EPJ Web of Conferences*, vol. 146, (Les Ulis, France: EDP Sciences, 2017), 09037; L. El-Guebaly, V. Massaut, K. Tobita, and L. Cadwallader, "Goals, Challenges, and Successes of Managing Fusion Activated Materials," *Fusion Engineering and Design* 83 (2008): 928–35.

20. R. Conn et al., "Economic, Safety and Environmental Prospects of Fusion Reactors," *Nuclear Fusion* 30 (1990): 1919.

21. B. K. Sovacool et al., "Balancing Safety with Sustainability: Assessing the Risk of Accidents for Modern Low-Carbon Energy Systems," *Journal of Cleaner Production* 112 (2016): 3952–965; S. Gordelier, *Comparing Nuclear Accident Risks with Those from Other Energy Sources* (OECD, 2010), doi:http://dx.doi.org/10.1787/9789264097995-en; "Deaths per TWh by Energy Source," NextBigFuture.com (2011), https://www.nextbigfuture.com/2011/03/deaths-per-twh-by-energy-source.html; A. Markandya and P. Wilkinson, "Electricity Generation and Health," *Lancet* 370 (2007): 979–90.

22. F. Richter, M. Steenbeck, M. Wilhelm, et al., *Nuclear Accidents and Policy: Notes on Public Perception* (DIW Berlin, the German Socio-Economic Panel [SOEP], 2013); P. A. Kharecha and M. Sato, "Implications of Energy and CO_2 Emission Changes in Japan and Germany After the Fukushima Accident," *Energy Policy* 132 (2019): 647–53; M. J. Neidell, S. Uchida, and M. Veronesi, *Be Cautious with the Precautionary Principle: Evidence from Fukushima Daiichi Nuclear Accident* (Cambridge, MA: National Bureau of Economic Research, 2019).

23. P. A. Kharecha and J. E. Hansen, "Prevented Mortality and Greenhouse Gas Emissions from Historical and Projected Nuclear Power," *Environmental Science & Technology* 47 (2013): 4889–895.

Chapter 9: Finishing the Race for Fusion

1. E. Lawrence, Ernest Lawrence banquet speech, the Nobel Foundation, 1940 (Nobel Media AB, 2020), https://www.nobelprize.org/prizes/physics/1939/lawrence/speech/.

2. D. van Houtte et al., "Recent Fully Non-Inductive Operation Results in Tore Supra with 6 Min, 1GJ Plasma Discharges," *Nuclear Fusion* 44 (2004): L11–L15; X. Gong et al., "Integrated Operation of Steady-State Long-Pulse H-Mode in Experimental Advanced Superconducting Tokamak," *Nuclear Fusion* 59 (2019): 086030; Phys Org, "Korean Artificial Sun sets the New World Record of 20-Sec-Long Operation at 100 Million Degrees" (2020), https://phys.org/news/2020-12-korean-artificial-sun-world-sec-long.amp.

3. J. Wesson and D. J. Campbell, *Tokamaks*, vol. 149 (Oxford: Oxford University Press, 2011); F. Wagner et al., "Development of an Edge Transport Barrier at the H Mode Transition of ASDEX," *Physical Review Letters* 53 (1984): 1453–456;

F. Wagner, "A Quarter-Century of H-Mode Studies," *Plasma Physics and Controlled Fusion* 49 (2007): B1–B33; R. Arnoux, "How Fritz Wagner 'Discovered' the H-Mode," *Iter Newsline* 86 (2009), https://www.iter.org/newsline/86/659; "Thirty Years of H-Mode," EUROfusion.org (2012), https://www.euro-fusion.org/news/detail/thirty-years-of-h-mode/?.

4. J. Kates-Harbeck, A. Svyatkovskiy, and W. Tang, "Predicting Disruptive Instabilities in Controlled Fusion Plasmas Through Deep Learning," *Nature* 568 (2019): 526; G. Kluth et al., "Deep Learning for NLTE Spectral Opacities," *Physics of Plasmas* 27 (2020): 052707.

5. T. Boisson, "British Nuclear Fusion Reactor Relaunched for the First Time in 23 Years," *Trust My Science* (2020), https://trustmyscience.com/reacteur-fusion-anglais-relance-premiere-fois-depuis-23-ans/.

6. J. Wesson and D. J. Campbell, *Tokamaks*, vol. 149 (Oxford: Oxford University Press, 2011); A. E. Costley, "On the Fusion Triple Product and Fusion Power Gain of Tokamak Pilot Plants and Reactors," *Nuclear Fusion* 56 (2016): 066003.

7. T. Fujita et al., "High Performance Experiments in JT-60U Reversed Shear Discharges," *Nuclear Fusion* 39 (1999): 1627–636;"Wendelstein 7-X Achieves World Record," Max Planck Institute for Plasma Physics (2018), https://www.ipp.mpg.de/4413312/04_18; T. S. Pedersen et al., "First Results from Divertor Operation in Wendelstein 7-X," *Plasma Physics and Controlled Fusion* 61 (2018): 014035.

8. P. O'Shea, M. Laberge, M. Donaldson, M. Delage, et al., "Acoustically Driven Magnetized Target Fusion at General Fusion: An Overview," *Bulletin of the American Physical Society* 61 (2016); D. Clery, "Alternatives to Tokamaks: A Faster-Better-Cheaper Route to Fusion Energy?," *Philosophical Transactions of the Royal Society A: Mathematical, Physical, and Engineering Sciences* 377 (2019): 20170431; R. Mumgard, *A New Approach to Funding, Accelerating, and Commercializing Fusion: NAS Comments, PPPL* (Commonwealth Fusion Systems, 2018).

9. J. Wesson and D. J. Campbell, *Tokamaks*, vol. 149 (Oxford: Oxford University Press, 2011).

10. M. Claessens, *ITER: The Giant Fusion Reactor: Bringing a Sun to Earth* (London: Springer Nature, 2019).

11. "ITER FAQ" (2020), http://www.iter.org/faq; E. Cartlidge, "Fusion Energy Pushed Back Beyond 2050," BBC (2017), https://www.bbc.co.uk/news/science-environment-40558758.

12. ITER Organisation, *ITER Research Plan Within the Staged Approach (Level III—Provisional Version)*, ITER (2018); G. Brennan, "When Will Fusion Power European Grids?—the Commercial Reactor—Part 2," *Engineers Journal* (2016), http://www.engineersjournal.ie/2016/02/09/when-will-fusion-power-european-grids-the-commercial-reactor-part-2/.

13. US Department of Energy, *2015 Review of the Inertial Confinement Fusion and High Energy Density Science Portfolio* (National Nuclear Security Administration, 2016).

14. O. A. Hurricane et al., "Fuel Gain Exceeding Unity in an Inertially Confined Fusion Implosion," *Nature* 506 (2014): 343–48; O. Hurricane et al., "Approach-

ing a Burning Plasma on the NIF," *Physics of Plasmas* 26 (2019): 052704; S. Le Pape et al., "Fusion Energy Output Greater Than the Kinetic Energy of an Imploding Shell at the National Ignition Facility," *Physical Review Letters* 120 (2018): 245003; D. Clery, "Laser Fusion Reactor Approaches 'Burning Plasma' Milestone," *Science* 370 (2020): 1019–20.

15. D. Clark et al., "Three-Dimensional Modeling and Hydrodynamic Scaling of National Ignition Facility Implosions," *Physics of Plasmas* 26 (2019): 050601; V. Gopalaswamy et al., "Tripled Yield in Direct-Drive Laser Fusion Through Statistical Modelling," *Nature* 565 (2019): 581–86.

16. K. Hahn et al., "Fusion-Neutron Measurements for Magnetized Liner Inertial Fusion Experiments on the Z Accelerator," in *Journal of Physics: Conference Series*, vol. 717 (IOP Publishing, 2016), 012020.

17. O. Hurricane et al., "Approaching a Burning Plasma on the NIF," *Physics of Plasmas* 26 (2019): 052704; P. Amendt et al., "Ultra-High (>30%) Coupling Efficiency Designs for Demonstrating Central Hot-Spot Ignition on the National Ignition Facility Using a Frustraum," *Physics of Plasmas* 26 (2019): 082707.

18. R. Aymar, P. Barabaschi, and Y. Shimomura, "The ITER Design," *Plasma Physics and Controlled Fusion* 44 (2002): 519–65.

19. M. Claessens, *ITER: The Giant Fusion Reactor: Bringing a Sun to Earth* (London: Springer Nature, 2019); H. A. Bethe, "The Fusion Hybrid," *Nuclear News* 21 (1978): 41; "Russia Develops a Fission-Fusion Hybrid Reactor," *Nuclear Engineering International Magazine* (2018), https://www.neimagazine.com/news/newsrussia-develops-a-fission-fusion-hybrid-reactor-6168535; R. Barrett and R. Hardie, *Fusion-Fission Hybrid as an Alternative to the Fast Breeder Reactor* (Los Alamos Scientific Lab, 1980).

20. T. Klinger et al., "Overview of First Wendelstein 7-X High-Performance Operation," *Nuclear Fusion* 59 (2019): 112004; F. Warmer et al., "From W7-X to a HELIAS Fusion Power Plant: On Engineering Considerations for Next-Step Stellarator Devices," *Fusion Engineering and Design* 123 (2017): 47–53.

21. I. T. Chapman and A. Morris, "UKAEA Capabilities to Address the Challenges on the Path to Delivering Fusion Power," *Philosophical Transactions of the Royal Society A: Mathematical, Physical and Engineering Sciences* 377 (2019): 20170436; T. Tanabe et al., "Tritium Retention of Plasma Facing Components in Tokamaks," *Journal of Nuclear Materials* 313 (2003): 478–90.

22. I. T. Chapman and A. Morris, "UKAEA Capabilities to Address the Challenges on the Path to Delivering Fusion Power," *Philosophical Transactions of the Royal Society A: Mathematical, Physical and Engineering Sciences* 377 (2019): 20170436; S. Brezinsek et al., "Fuel Retention Studies with the ITER-like Wall in JET," *Nuclear Fusion* 53 (2013): 083023; A. Baron-Wiechec et al., "First Dust Study in JET with the ITER-like Wall: Sampling, Analysis and Classification," *Nuclear Fusion* 55 (2015): 113033.

23. M. Claessens, *ITER: The Giant Fusion Reactor: Bringing a Sun to Earth* (London: Springer Nature, 2019); A. Donné, "The European Roadmap Towards Fusion Electricity," *Philosophical Transactions of the Royal Society A: Mathematical, Physical and Engineering Sciences* 377 (2019): 20170432.

24. R. Miles et al., "Thermal and Structural Issues of Target Injection into a Laser-

Driven Inertial Fusion Energy Chamber," *Fusion Science and Technology* 66 (2014): 343–48.

25. P. Mason et al., "Kilowatt Average Power 100 J-Level Diode Pumped Solid State Laser," *Optica* 4 (2017): 438.

26. W. Meier et al., "Fusion Technology Aspects of Laser Inertial Fusion Energy (Life)," *Fusion Engineering and Design* 89 (2014): 2489–492; M. Dunne et al., "Timely Delivery of Laser Inertial Fusion Energy (LIFE)," *Fusion Science and Technology* 60 (2011): 19–27; T. M. Anklam, M. Dunne, W. R. Meier, S. Powers, and A. J. Simon, "LIFE: The Case for Early Commercialization of Fusion Energy," *Fusion Science and Technology* 60 (2011): 66–71.

27. D. Clery, "Knighthood in Hand, Astrophysicist Prepares to Lead U.S. Fusion Lab," *Science* (2019), https://www.sciencemag.org/news/2018/06/knighthood -hand-astrophysicist-prepares-lead-us-fusion-lab.

28. D. Clery, *A Piece of the Sun: The Quest for Fusion Energy* (New York: Abrams, 2014); N. R. Council, S. E. Koonin, et al., *Second Review of the Department of Energy's Inertial Confinement Fusion Program* (Washington, DC: National Academies Press, 1990); D. N. Bixler, "The LMF: Riding Out the Tide of Change," *Journal of Fusion Energy* 10 (1991): 335–37; "National Ignition Facility. FAQs," Lawrence Livermore National Laboratory (2020), https://lasers.llnl.gov/about /faqs; N. R. Council et al., *Review of the Department of Energy's Inertial Confinement Fusion Program: The National Ignition Facility* (Washington, DC: National Academies Press, 1997).

29. A. Sykes et al., "Compact Fusion Energy Based on the Spherical Tokamak," *Nuclear Fusion* 58 (2017): 016039; A. E. Costley, "On the Fusion Triple Product and Fusion Power Gain of Tokamak Pilot Plants and Reactors," *Nuclear Fusion* 56 (2016): 066003; J. D. Farmer and F. Lafond, "How Predictable Is Technological Progress?," *Research Policy* 45 (2016): 647–65; H. Ritchie, "Renewable Energy," *Our World in Data* (2017), https://ourworldindata.org/renewable -energy.

30. D. Castelvecchi, "Next-Generation LHC: CERN Lays Out Plans for €21-Billion Supercollider," *Nature* (2019), https://www.nature.com/articles/d41586-019 -00173-2; E. Gibney and D. Castelvecchi, "CERN Makes Bold Push to Build €21-Billion Supercollider," *Nature* (2020), https://www.nature.com/articles /d41586-020-01866-9; A. Knapp, "How Much Does It Cost to Find a Higgs Boson?," *Forbes* (2012), https://www.forbes.com/sites/alexknapp/2012/07/05 /how-much-does-it-cost-to-find-a-higgs-boson/#28829a2c3948; J. R. Minkel, "Is the International Space Station Worth $100 billion?," Space.com (2010), https://www.space.com/9435-international-space-station-worth-100-billion .html; E. Cartlidge, "Square Kilometre Array Hit with Further Cost Hike and Delay," *Physics World* (2019), https://physicsworld.com/a/square-kilometre -array-hit-with-further-cost-hike-and-delay/.

31. "Total Energy Price and Expenditure Estimates (Total, per Capita, and per GDP), Ranked by State, 2018," US Energy Information Administration (2020), https://www.eia.gov/state/seds/data.php?incfile=/state/seds/sep_sum/html /rank_pr.html&sid=US.

32. S. O. Dean, "Historical Perspective on the United States Fusion Program," *Fu-

sion Science and Technology 47 (2005): 291–99; S. O. Dean, "Fusion Power by Magnetic Confinement Program Plan," *Journal of Fusion Energy* 17 (1998): 263–87; "Gross Domestic Spending on R&D," OECD iLibrary (2020), doi:10.1787/d8b068b4-en; R. E. Rowberg, "Congress and the Fusion Energy Sciences Program: A Historical Analysis," *Journal of Fusion Energy* 18 (1999): 29–46.

33. "Gross Domestic Spending on R&D," OECD iLibrary (2020), doi:10.1787/d8b068b4-en; "Federal Science Budget Tracker," American Institute of Physics (2020), https://www.aip.org/fyi/federal-science-budget-tracker/FY2020.

34. S. Eckhouse, G. Lewison, and R. Sullivan, "Trends in the Global Funding and Activity of Cancer Research," *Molecular Oncology* 2 (2008): 20–32.

35. N. Hawker, "A Simplified Economic Model for Inertial Fusion," *Philosophical Transactions of the Royal Society A: Mathematical, Physical and Engineering Sciences* 378 (2020): 20200053.

36. T. M. Anklam, M. Dunne, W. R. Meier, S. Powers, and A. J. Simon, "LIFE: The Case for Early Commercialization of Fusion Energy," *Fusion Science and Technology* 60 (2011): 66–71; Bechtel National, Inc., *Fusion Power Capital Cost Study* (ARPA-E, 2017).

37. Energy Information Administration, *Levelized Cost and Levelized Avoided Cost of New Generation Resources in the Annual Energy Outlook* (US Government, 2019); IEA, *Projected Costs of Generating Electricity 2020* (2020), https://www.iea.org/reports/projected-costs-of-generating-electricity-2020.

38. L. L. Strauss, "Remarks Prepared for Delivery at the Founders Day Dinner," *National Association of Science Writers* 16 (1954).

Epilogue—Can We Afford Not to Do Fusion?

1. L. Artsimovich, "Matter and Energy," in *Children's Encyclopedia* (ed. Alexei Ivanovich Markushevich) (Pedagogy, 1973).

2. R. Black, "The Top 10 Greatest Survivors of Evolution," *Smithsonian* (2012), http://www.smithsonianmag.com/science-nature/The-Top-10-Greatest-Survivors-of-Evolution-178186561.html; D. M. Raup and S. J. Gould, *Extinction: Bad Genes or Bad Luck?* (New York: W. W. Norton & Company, 1993); D. Jablonski and W. G. Chaloner, "Extinctions in the Fossil Record [and Discussion]," *Philosophical Transactions of the Royal Society of London B: Biological Sciences* 344 (1994): 11–17; C. Lavett Smith, C. S. Rand, B. Schaeffer, and J. W. Atz, "Latimeria, the Living Coelacanth, Is Ovoviviparous," *Science* 190 (1975): 1105–1106.

3. C. R. Chapman and D. Morrison, "Impacts on the Earth by Asteroids and Comets: Assessing the Hazard," *Nature* 367 (1994): 33–40; Z. Sekanina, "The Tunguska Event—No Cometary Signature in Evidence," *Astronomical Journal* 88 (1983): 1382–413; S. Self, "The Effects and Consequences of Very Large Explosive Volcanic Eruptions," *Philosophical Transactions of the Royal Society A: Mathematical, Physical and Engineering Sciences* 364 (2006): 2073–97.

4. P. W. Anderson, "More Is Different," *Science* 177 (1972): 393–96.

5. G. Dosi, M. Napoletano, A. Roventini, J. E. Stiglitz, and T. Treibich, *Rational*

Heuristics? Expectations and Behaviors in Evolving Economies with Hetero-geneous Interacting Agents (Cambridge, MA: National Bureau of Economic Research, 2020), http://www.nber.org/papers/w26922, doi:10.3386/w26922; Y. Achdou, J. Han, J. M. Lasry, P. L. Lions, and B. Moll, *Income and Wealth Distribution in Macroeconomics: A Continuous-Time Approach* (Cambridge, MA: National Bureau of Economic Research, 2017), http://www.nber.org/papers/w23732, doi:10.3386/w23732.

6. J. Wolf, G. R. Asrar, and T. O. West, "Revised Methane Emissions Factors and Spatially Distributed Annual Carbon Fluxes for Global Livestock," *Carbon Balance and Management* 12 (2017): 16; M. Kamo, Y. Sato, S. Matsumoto, and N. Setaka, "Diamond Synthesis from Gas Phase in Microwave Plasma," *Journal of Crystal Growth* 62 (1983): 642–44.

7. S. V. Bulanov et al., "Laser Ion Acceleration for Hadron Therapy," *Physics-Uspekhi* 57 (2014): 1149–1179.

8. D. T. Casey et al., "Thermonuclear Reactions Probed at Stellar-Core Conditions with Laser-Based Inertial-Confinement Fusion," *Nature Physics* 13 (2017): 1227–231.

9. G. Wilt, "Glimpses of an Exceptional Man," *Science & Technology Review* (July/August, 1998).

10. P. M. Celliers et al., "Insulator-Metal Transition in Dense Fluid Deuterium," *Science* 361 (2018): 677–82.

11. I. T. Chapman and A. W. Morris, "UKAEA Capabilities to Address the Challenges on the Path to Delivering Fusion Power," *Philosophical Transactions of the Royal Society A: Mathematical, Physical and Engineering Sciences* 377 (2019): 20170436.

12. G. Wurden et al., "A New Vision for Fusion Energy Research: Fusion Rocket Engines for Planetary Defense," *Journal of Fusion Energy* 35 (2016): 123–33.

13. F. J. Dyson, "Interstellar Transport," *Physics Today* 21 (1968): 41–45.

14. D. B. Lombard, "Plowshare: A Program for the Peaceful Uses of Nuclear Explosives," *Physics Today* 14 (1961): 24–34; G. W. Johnson and H. Brown, "Non-Military Uses of Nuclear Explosives," *Scientific American* 199 (1958): 29–35; E. Teller, *Plowshare* (Livermore, CA: University of California, 1963); C. R. Gerber, R. Hamburger, and E. S. Hull, *Plowshare* (Washington, DC: US Atomic Energy Commission, Division of Technical Information, 1967); M. D. Nordyke, "The Soviet Program for Peaceful Uses of Nuclear Explosions," *Science & Global Security* 7 (1998): 1–117.

15. J. Cassibry et al., "Case and Development Path for Fusion Propulsion," *Journal of Spacecraft and Rockets* 52 (2015): 595–612; G. Schmidt, J. Bonometti, and P. Morton, "Nuclear Pulse Propulsion—Orion and Beyond," in *36th AIAA/ASME/SAE/ASEE Joint Propulsion Conference and Exhibit* (2000): 3856; C. Orth, *Interplanetary Space Transport Using Inertial Fusion Propulsion* (Lawrence Livermore National Lab, 1998); I. A. Crawford, "Interstellar Travel: A Review for Astronomers," *Quarterly Journal of the Royal Astronomical Society* 31 (1990): 377–400; K. Long et al., "PROJECT ICARUS: Son of Daedalus, Flying Closer to Another Star," *arXiv preprint arXiv:1005.3833* (2010); W. Moeckel, "Comparison of Advanced Propulsion Concepts for Deep Space Exploration," *Journal of Spacecraft and Rockets* 9 (1972): 863–68.

INDEX

INDEX

INDEX